Nanotechnology
Consequences for Human Health and the Environment

ISSUES IN ENVIRONMENTAL SCIENCE AND TECHNOLOGY

EDITORS:

TITLES IN THE SERIES:

How to obtain future titles on publication

A subscription is available for this series. This will bring delivery of each new volume immediately on publication and also provide you with online access to each title via the Internet. For further information visit http://www.rsc.org/Publishing/Books/issues or write to the address below.

For further information please contact:
Sales and Customer Care, Royal Society of Chemistry, Thomas Graham House, Science Park, Milton Road, Cambridge, CB4 0WF, UK
Telephone: +44 (0)1223 432360, Fax: +44 (0)1223 426017, Email: sales@rsc.org

ISSUES IN ENVIRONMENTAL SCIENCE AND TECHNOLOGY

EDITORS: R.E. HESTER AND R.M. HARRISON

24
Nanotechnology:
Consequences for Human Health and the Environment

RSCPublishing

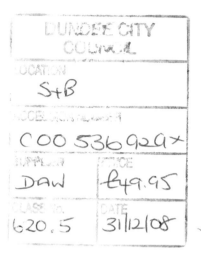
ISBN-13: 978-0-85404-216-6
ISSN: 1350-7583

A catalogue record for this book is available from the British Library

Published by The Royal Society of Chemistry,
Thomas Graham House, Science Park, Milton Road,
Cambridge CB4 0WF, UK

Registered Charity Number 207890

For further information see our web site at www.rsc.org

Preface

Few outside of the world of science and technology have much concept of what nanotechnology involves. It is defined in terms of products and processes involving nanometre (*i.e.* 10^{-9} or 0.000 000 001 m) dimensions but this gives no flavour for what is truly involved. What may be surprising to many is that there is a massive thrust of research and development leading to new products involving nanoscale materials and it is projected that this will be a multi-billion dollar industry within a matter of a few years. Having in the past failed to anticipate the adverse public health consequences of products such as asbestos, governments around the world are investing resource into assessing the possible adverse consequences arising from the present and future application of nanotechnologies. This led the Royal Society and the Royal Academy of Engineering in the UK to publish an expert report on the topic under the title of "Nanoscience and nanotechnologies: opportunities and uncertainties". One manifestation of this government's concern is that in the UK a system has been introduced by the government for the voluntary notification of products and processes using nanoscale materials.

Some nanoscale materials such as carbon black, titanium dioxide and silica have been in high tonnage production in industry for many years, with a wide range of uses. However, a vast range of other nanoscale materials are now being produced with uses as diverse as manufacturing tennis balls which retain their bounce for longer and underwear with an antimicrobial coating. The concerns over nanoparticles and nanotubes relate to the observation that they are more toxic per unit mass than the same materials in larger particle forms. Whilst the evidence for extreme toxicity of the traditionally produced nanoscale materials is lacking, there remains concern that new forms of engineered nanomaterials may prove to be appreciably toxic. There is no doubt that by virtue of their size they have a much stronger ability to penetrate into the human body than more conventionally sized materials.

This volume of Issues seeks to give a broad overview of the sources, behaviour and risks associated with nanotechnology. In the first chapter, Barry Park of Oxonica Limited, a company specialising in nanoscale products, gives an overview of the current and future applications of nanotechnology. This is followed by a discussion of nanoparticles in the aquatic and terrestrial environment by Jamie Lead of the University of Birmingham, which includes consideration of the behaviour of nanoparticles both in the aquatic environment and within soils where they can be used in remediation processes. This is followed in a third chapter by Roy Harrison with a consideration of nanoparticles within the atmosphere. Currently, this is the most important medium for human exposure, although there is very limited evidence that nanoparticles play a particularly prominent role within the overall toxicity of airborne particulate matter.

Currently, those receiving the highest exposures to nanoparticles and nano-tubes are those people occupationally exposed in the industry, and in the following chapter David Mark of the Health and Safety Laboratory describes the issues of occupational exposure, including how it can be assessed and currently available data from industrial sites. The following two chapters deal respectively with the toxicological properties and human health effects of nanoparticles. In the former chapter, Ken Donaldson and Vicki Stone give a toxicological perspective on the properties of nanoparticles and consider why nanoparticle form may confer an especially high level of toxicity. This is then put into context in the following chapter by Lang Tran and co-authors, which looks for hard evidence of adverse effects upon human health both in the occupational environment and in outside air.

This volume is rounded off by a chapter by Andrew Maynard, Chief Science Adviser to the Project on Emerging Nanotechnologies of the Woodrow Wilson International Center for Scholars in the United States, which highlights the problems of regulation that are presented by a burgeoning nanotechnology industry and gives some comfort in that the problems and solutions emerging in North America do not differ greatly from those being formulated within Europe.

Overall, the volume provides a comprehensive overview of the current issues concerning engineered nanoparticles which we believe will be of immediate value to scientists, engineers and policymakers within the field, as well as to students on advanced courses wishing to look closely into this topical subject.

Ronald E. Hester
Roy M. Harrison

Contents

Current and Future Applications of Nanotechnology
Barry Park

Nanoparticles in the Aquatic and Terrestrial Environments
Jamie Lead

Nanoparticles in the Atmosphere
Roy Harrison

Occupational Exposure to Nanoparticles and Nanotubes
David Mark

Toxicological Properties of Nanoparticles and Nanotubes
Ken Donaldson and Vicki Stone

Human Effects of Nanoparticle Exposure
Lang Tran, Rob Aitken, Jon Ayres, Ken Donaldson and Fintan Hurley

Nanoparticle Safety – A Perspective from the United States
Andrew D. Maynard

Editors

Ronald E. Hester, BSc, DSc(London), PhD(Cornell), FRSC, CChem

Ronald E. Hester is now Emeritus Professor of Chemistry in the University of York. He was for short periods a research fellow in Cambridge and an assistant professor at Cornell before being appointed to a lectureship in chemistry in York in 1965. He was a full professor in York from 1983 to 2001. His more than 300 publications are mainly in the area of vibrational spectroscopy, latterly focusing on time-resolved studies of photoreaction intermediates and on biomolecular systems in solution. He is active in environmental chemistry and is a founder member and former chairman of the Environment Group of the Royal Society of Chemistry and editor of 'Industry and the Environment in Perspective' (RSC, 1983) and 'Understanding Our Environment' (RSC, 1986). As a member of the Council of the UK Science and Engineering Research Council and several of its sub-committees, panels and boards, he has been heavily involved in national science policy and administration. He was, from 1991 to 1993, a member of the UK Department of the Environment Advisory Committee on Hazardous Substances and from 1995 to 2000 was a member of the Publications and Information Board of the Royal Society of Chemistry.

Roy M. Harrison, BSc, PhD, DSc(Birmingham), FRSC, CChem, FRMetS, Hon MFPH, Hon FFOM

Roy M. Harrison is Queen Elizabeth II Birmingham Centenary Professor of Environmental Health in the University of Birmingham. He was previously Lecturer in Environmental Sciences at the University of Lancaster and Reader and Director of the Institute of Aerosol Science at the University of Essex. His more than 300 publications are mainly in the field of environmental chemistry, although his current work includes studies of human health impacts of atmospheric pollutants as well as research into the chemistry of pollution phenomena. He is a past Chairman of the Environment Group of the Royal Society of Chemistry for whom he has edited 'Pollution: Causes, Effects and Control' (RSC, 1983; Fourth Edition, 2001) and 'Understanding our Environment: An Introduction to Environmental Chemistry and Pollution' (RSC, Third Edition, 1999). He has a close interest in scientific and policy aspects of air pollution, having been Chairman of the Department of Environment Quality of Urban Air Review Group and the DETR Atmospheric Particles Expert Group as well as a member of the Department of Health Committee on the Medical Effects of Air Pollutants. He is currently a member of the DEFRA Air Quality Expert Group, the DEFRA Advisory Committee on Hazardous Substances and the DEFRA Expert Panel on Air Quality Standards.

Contributors

Rob Aitken, Institute of Occupational Medicine, Research Avenue North, Riccarton, Edinburgh, EH14 4AP, Scotland, UK

Jon Ayres, Liberty Safe Work Research Centre, Foresterhill Road, Aberdeen AB25 2ZP, Scotland, UK

Ken Donaldson, MRC/University of Edinburgh Centre for Inflammation Research, ELEGI Colt Laboratory, Queen's Medical Research Institute, 47 Little France Crescent, Edinburgh, EH16 4TJ, Scotland, UK

Roy Harrison, Division of Environmental Health & Risk Management, School of Geography, Earth & Environmental Sciences, University of Birmingham, Edgbaston, Birmingham B15 2TT, UK

Fintan Hurley, Institute of Occupational Medicine, Research Avenue North, Riccarton, Edinburgh, EH14 4AP, Scotland, UK

Jamie Lead, Division of Environmental Health & Risk Management, School of Geography, Earth & Environmental Sciences, University of Birmingham, Edgbaston, Birmingham B15 2TT, England, UK

David Mark, Health and Safety Laboratory, Harpur Hill, Buxton, Derbyshire, SK17 9JN, England, UK

Andrew Maynard, Wilson International Center for Scholars, One Woodrow Wilson Plaza, 1300 Pennsylvania Ave., NW Washington, DC 20004-3027, USA

Barry Park, Oxonica Limited, 7 Begbroke Science Park, Sandy Lane, Yarnton, Kidlington, Oxfordshire, OX5 1PF, England, UK

Vicki Stone, Centre for Health and Environment, School of Life Sciences, Napier University, Merchiston Campus, Edinburgh, EH10 5DT, Scotland, UK

Lang Tran, Institute of Occupational Medicine, Research Avenue North, Riccarton, Edinburgh, EH14 4AP, Scotland, UK

Current and Future Applications of Nanotechnology

BARRY PARK

1 Introduction

1.1 History

Physicist Richard P. Feynman first described the concept of nanoscience in 1959 in a lecture to the American Physical Society and the term nanotechnology was coined in 1974 by the Japanese researcher Norio Taniguchi[1] to describe precision engineering with tolerances of a micron or less. In the mid 1980s, Eric Drexler brought nanotechnology into the public domain with his book *Engines of Creation*.[2]

1.2 Definitions

As part of a major report commissioned by the UK Government from the Royal Society and the Royal Academy of Engineering in the UK, entitled "Nanoscience and nanotechnologies: opportunities and uncertainties",[3] the following definitions were used:

Nanoscience is the study of phenomena and manipulation of materials at atomic, molecular and macromolecular scales, where properties differ significantly from those at a larger scale.

Nanotechnologies are the design, characterisation, production and application of structures, devices and systems by controlling shape and size at nanometre scale.

The NASA website provides an interesting definition of nanotechnology: "The creation of functional materials, devices and systems through control of matter on the nanometre scale (1–100 nm) and exploitation of novel phenomena and properties (physical, chemical, biological) at that length scale."[4]

Issues in Environmental Science and Technology, No. 24
Nanotechnology: Consequences for Human Health and the Environment

The Oxford English Dictionary defines nanotechnology as "technology on an atomic scale, concerned with dimensions of less than 100 nanometres".

The prefix *nano-* derives from the Greek word for dwarf and one nanometre is equal to one billionth of a metre *i.e.* 10^{-9} m. Nanomaterials are therefore regarded as those that have at least one dimension of size less than 100 nm.

1.3 Investment

Nanotechnology has received very significant investment over the past ten years with national governments providing the bulk of this investment with estimates ranging as high as $18 billion for investment between 1997 and 2005.[5] There has recently been a four-way split with similar investment in each of USA, Europe, Japan and the rest of the world with approximately $3 billion spent by governments in 2003 alone.[6] In the USA, for example, the National Nanotechnology Initiative (NNI) is a federal R&D program to coordinate the multi-agency efforts in nanoscale science, engineering and technology.

The President's 2007 budget provides over $1.2 billion for the Initiative, bringing the total investment since the NNI was established in 2001 to over $6.5 billion and nearly tripling the annual investment of the first year of the Initiative.[7] With this investment has come a large number of products, some of which are already on the market, that are based on nanotechnology or contain nanomaterials.

2 Technology

2.1 Nanomaterials

There had already been exploitation of products of particle size falling within the definition of a nanomaterial prior to these developments, but the products were simply referred to as ultrafine or superfine. These products, mainly comprising metal or metalloid oxides and carbon blacks, were primarily additives for the plastics industry in its various guises and these will be considered in some detail as they comprise the greatest body of current applications of nanotechnology. Alongside these products that have considerable sales value are many novel products, which are currently available from a range of new companies and generally started from work originating from research studies in a university. Applications of these products are wide and again these will be considered.

Nanomaterials can be considered under the following three headings:

(i) Natural
(ii) Anthropogenic (adventitious)
(iii) Engineered

Natural nanomaterials comprise those created independently of man and include a wide range of materials that contain a nanocomponent and may be found in the atmosphere such as sea salt resulting from the evaporation of water from sea spray, soil dust, volcanic dust, sulfates from biogenic gases, organics from biogenic gases and nitrates from NO_x. The actual content of any one or a combination of these nanomaterials in the atmosphere is dependent on geography.

Anthropogenic (adventitious) nanomaterials are those created as a result of action by man with the main example of this type of nanomaterial being soot resulting from the combustion of fossil fuels. Other anthropogenic nano-materials include welding fume and particulates resulting from the oxidation of gases such as sulfates and nitrates.

These two types of nanomaterials comprise many examples, some of which have been studied in great depth especially to minimise damage to health from exposure to these materials.

The subject of this paper falls largely in the third category, *i.e.* engineered nanomaterials, which have been designed and manufactured by man. These have been synthesised for a specific purpose and may be found in one of several different shapes. As defined above, the term nano describes the size in at least one dimension so nanomaterials may have nano characteristics in one, two or three dimensions. These correspond to platelet-like, wire-like and spheroidal structures respectively. The engineered nanomaterials may be further sub-divided into organic and inorganic types, with the former including carbon itself and polymeric structures with specific nano characteristics. Inorganics include metals, metal and metalloid oxides, clays and a specific subset of compounds known as quantum dots.

2.2 Manufacturing Processes

Nanoparticles can be produced by a variety of methods. These include com-bustion synthesis, plasma synthesis, wet-phase processing, chemical precipita-tion, sol-gel processing, mechanical processing, mechanicochemical synthesis, high-energy ball-milling, chemical vapour deposition and laser ablation.

2.3 Product Characteristics

In summary, the key characteristics of nanomaterials that define their potential applications include the following:

- High surface area
- High activity
- Catalytic surface
- Adsorbent
- Prone to agglomeration
- Range of chemistries
- Natural and synthetic
- Wide range of applications

3   Types of Nanomaterials

3.1   Carbon

3.1.1   Carbon Black.   Carbon black accounts for the largest tonnage of engineered nanomaterial and carbon blacks are used in a wide variety of applications, including printing inks, toners, coatings, plastics, paper and building products. Dependent on the size and chemistry of the particles, carbon-black-containing plastics can be electrically conducting or insulating and have significant reinforcing characteristics.[8,9]

Carbon black is a very fine particulate form of elemental carbon and was first produced more than 2000 years ago by the ancient Chinese and Egyptians for use as a colourant.[10] Although carbon black is still valued today for its colouring attributes, it is primarily used to provide reinforcement and other properties, especially to rubber articles. All carbon black is produced either by incomplete combustion or thermal decomposition of a hydrocarbon feedstock.

Two important characteristics of carbon black are surface area, an indirect measure of particle size, and structure, a measure of the degree of particle aggregation or chaining. Surface areas of carbon blacks can range from *c.* 10 m^2 g^{-1}, for use as reinforcing fillers, up to *c.* 1100 m^2 g^{-1}, for use as electrically conductive fillers. Surface area and structure are dependent on the type of process to manufacture the carbon black and they define the performance of the carbon black in its application.

The mass production of carbon blacks started in the first half of the twentieth century in the wake of the expanding tyre industry. Carbon blacks were used as reinforcing fillers to optimise the physical properties of tyres and make them more durable. Even today the tyre industry uses at least 70% of the carbon blacks manufactured worldwide. The remainder finds use in a range of applications. Carbon blacks are now widely used for plastics masterbatch applications for use in conductive packaging, films, fibres, mouldings, pipes and semiconductive cable jackets. They are also used as toners for printers and in printing inks. Carbon blacks can provide pigmentation, conductivity and UV protection for a number of coating applications including marine, aerospace and industrial. In at least some of these applications the coating requires UV curing and specific formulations have to be employed to overcome the inherent UV protection given by the carbon black during this process.[11,12]

The global market for carbon blacks is forecast to rise 4% per year through 2008 to 9.6 million metric tonnes.[13] The smaller non-tyre segment will show strongest gains. This segment also commands the highest prices with applications such as conductive fillers showing greatest growth prospects. Applications for plastics containing conductive fillers include antistatic surfaces and coatings.

3.1.2   Graphite.   One-dimensional carbon is classically graphite, which has sub-nano thickness layers and nano-size spacing between layers leading to use as a lubricant, where advantage can be taken of the ability of these layers to slide across one another reducing friction between two surfaces coated with this material. This spacing is being considered for use as a hydrogen store with potential application in hydrogen fuel cells. Mono-layer graphite, or graphene, has been demonstrated as having novel magnetic properties.

Graphene has a unique electronic structure and theory suggests that novel magnetic properties may be dependent on this structure. The graphene magnetic susceptibility is temperature dependent and increases with the amount of defects in the structure. Work has been done to confirm such novel properties although there has been no commercialisation of this property at present.[14]

Recent work has calculated that graphene spaced between 6 and 7 angstroms apart can store hydrogen at room temperature and moderate pressures. The amount of hydrogen stored comes close to a practical goal of 62 kg per cubic metre set by the US Department of Energy. Another advantage of this form of graphite is that the hydrogen gas can be released by moderate warming. The current challenge is to synthesise graphenes with the appropriate interplanar spacing for maximum hydrogen absorption. If this can be achieved then graphene could be a strong contender for practical hydrogen storage. It has been reported that "tuneable" graphite nanostructures could be created with different hydrogen storage properties by interposing space molecules between the graphite layers.[15,16] These spacers would have the added advantage of keeping out contaminants such as nitrogen and carbon monoxide, which can reduce hydrogen storage capacity.

3.1.3 Carbon Nanotubes. Carbon nanotubes are fullerene-related structures that consist of graphene cylinders closed at either end with caps containing pentagonal rings. They exhibit extraordinary strength and unique electrical properties and are efficient conductors of heat along their length. They exist in single-wall and multi-wall forms. They have been used as composite fibres in polymers and concrete to improve the mechanical, thermal and electrical properties of the bulk product. They have also been used as brushes for electrical motors. Inorganic variants have also been produced.

A nanotube is cylindrical with at least one end typically capped with a hemisphere of the buckyball structure. There are two main types of nanotube: single-wall nanotubes (SWNTs) and multi-wall nanotubes (MWNTs). Single-wall nanotubes have a diameter of *c.* 1 nm and a length that can be many thousands of times larger *i.e.* to the order of centimetres.[17] Single-wall nanotubes exhibit electric properties not shared by the multi-wall variants. They are therefore the most likely candidates for miniaturising electronics past the microelectromechanical scale that is currently the basis of modern electronics. The most basic building block of these systems is the electric wire and SWNTs can be excellent conductors.[18]

Carbon nanotubes are among the strongest materials known to man, in terms of both tensile strength and elastic modulus, and since carbon nanotubes have relatively low density, the strength to weight ratio is truly exceptional. They will bend to surprisingly large angles before they start to ripple and buckle and they finally develop kinks as well. These definitions are elastic, *i.e.* they all disappear completely when the load is removed.[19] They have already been used as composite fibres in polymers and concrete to improve the mechanical, thermal and electrical properties of the bulk product. Conductive carbon nanotubes have been used for several years in brushes for commercial electric motors. The carbon nanotubes permit reduced carbon in the brush.

Multi-wall nanotubes precisely nested within one another exhibit interesting properties whereby an inner nanotube may slide within its outer nanotube shell creating an atomically perfect linear or rotational bearing. This is one of the first true examples of molecular nanotechnology. Already this property has been utilised to create the world's smallest rotational motor and a rheostat. Future applications are likely to include conductive and high-strength composites, energy storage and energy conversion devices, sensors, field emission displays and radiation sources, hydrogen storage media, semiconductor devices, probes and interconnects.[20] Some of these are already products while others are in an early to advanced stage of development.[21]

3.1.4   Carbon "Buckyballs".   Fullerenes are the classic three-dimensional carbon nanomaterials. They have a unique structure comprising 60 carbon atoms in the shape reminiscent of a geodesic dome and are often referred to as "Buckyballs" or "Buckminsterfullerene", after the American architect R. Buckminster Fuller who designed the geodesic dome with the same fundamental symmetry. These C_{60} molecules comprise the same combination of hexagonal and pentagonal rings, and the name therefore has seemed appropriate. These spherical molecules were discovered in 1985 and considerable work has gone into their study. However, potential applications have been limited and include catalysts, drug delivery systems, optical devices, chemical sensors and chemical separation devices. The molecule can absorb hydrogen with enhanced absorption when transition metals are bound to the buckyballs, leading to potential use in hydrogen storage.[22,23]

3.2   Inorganic Nanotubes

Combinations of elements that can form stable two-dimensional sheets can be considered suitable to produce inorganic nanotubes and a number of inorganic chemists have been focusing on such structures.[24] Although the investment devoted to inorganic nanotubes lags behind that of carbon nanotubes, a number of reviews suggest that inorganic nanotube research is increasing rapidly.[25-27] Examples include tungsten sulfide[28] and boron nitride,[29] which may find uses where their inertness and high durability and conductivity can be exploited. Tungsten sulfide and molybdenum sulfide may have attractive lubricating properties.

Tenne was the first to report the synthesis of inorganic nanotubes[28] and has suggested a list of possible technologies that could use the unique properties of inorganic nanotubes. These include bullet-proof materials, high-performance sporting goods, specialised chemical sensors, catalysts and rechargeable batteries. As examples, titanium dioxide nanotubes have been shown to have potential as a hydrogen sensor[30] and in water photolysis.[31]

3.3 Metals

The simplest inorganic nanomaterials are metallic with a wide range of metals already produced in nano form. These include aluminium, copper, nickel, cobalt, iron, silver and gold with a wide range of potential applications including land remediation, batteries and explosives. Metal nanoparticles have been prepared for some time, but several have found significant commercial application. These include aluminium, iron, cobalt and silver.

3.3.1 Aluminium. Aluminium nanoparticles have been used for their pyrophoric characteristics in explosives.[32] Aluminium is a highly reactive metal when produced as a nanopowder and when in formulations such as metastable intermolecular composites (MIC) reacts to produce a large amount of heat energy. Aluminium powder is air stable due to a thin oxide shell that forms during production and protects the inner core from further oxidation.

3.3.2 Iron. Nanoscale iron particles have large surface areas and high surface reactivity and research has shown[32,33] that these particles are very effective for the transformation and detoxification of a wide variety of contaminants, such as chlorinated solvents, organochloric pesticides and polychlorinated biphenyls. Thus they have been used for remediation of soil and groundwater, which contains such contaminants.

3.3.3 Cobalt. Cobalt nanoparticles exhibit magnetic behaviour,[34-37] which may find application in medical imaging.[38]

3.3.4 Silver. Silver nanoparticles, which demonstrate antimicrobial and antibacterial activity,[39,40] have been used in a number of applications including medical dressings and non-smelling socks![41]

3.3.5 General. Special shaped metal nanometals hold promise for the miniaturisation of electronics, optics and sensors[42] where, for example, studies have shown that the conductance of copper nanowires is determined by the absorption of organic molecules.[43] Electrochemical deposition of palladium nanostructured films has led to potential application as calorimetric gas sensors for combustible gases.[44] In the biological sciences, many applications for metal nanoparticles are being explored, including biosensors,[39] labels for cells and biomolecules[45] and cancer therapeutics.[46]

3.4 Metal Oxides

The largest group of inorganic nanomaterials comprises metal oxides with titanium dioxide, zinc oxide and silicon dioxide as the largest volume materials.

Copper oxide, cerium oxide, zirconium oxide, aluminium oxide and nickel oxide have also been produced commercially and are available in bulk.

This category comprises the largest number of different types of nanomaterials. Conducting an internet search for nanomaterial manufacturers generates many hits, with most of the companies identified offering a range of metal oxide nanomaterials. These may or may not be currently produced in significant commercial quantities, but the manufacturing technology is generally capable of producing such materials in large quantities.

3.4.1 Titanium Dioxide. Titanium dioxide is used as a pigment in many applications including paints and paper with mean particle sizes of the order of 300 nm and accounts for approximately 4 000 000 tonnes per year. However, the existing market for ultrafine or nano titanium dioxide is about 4000 tonnes per year. The market for this material, whose mean particle size is in the range 20–80 nm, exploits the inherent strong scattering power in the UV while transmitting visible wavelengths through the crystal. The material in which ultrafine titanium dioxide is incorporated thus appears virtually transparent. Classically, the particles are coated with alumina, silica or zirconia or a combination of these oxides to ensure effective dispersion. Applications include products where protection of the substrate to the damaging rays of UV light is important. These include sunscreens, wood coatings, printing inks, paper and plastics. Rutile is the preferred crystal form of titanium dioxide for these applications, although anatase has also been used and is commercially available.

Nano or ultrafine titanium dioxide is available from a number of major manufacturers including Degussa, Kemira and Sachtleben in Europe and from ISK and Tayca in Japan.

Modified forms of titanium dioxide have also found markets. Oxonica has developed and is selling a manganese-doped titanium dioxide that exhibits significantly enhanced UVA absorption and minimises the generation of free radicals resulting from the absorption of UV light by the titanium dioxide.[47–49] This product is already being used commercially in sunscreens and cosmetics and is being evaluated for applications in coatings and plastics.

Doping titanium dioxide with tungsten or molybdenum produces a material that has enhanced photoactivity and Millennium produces nanoparticulate products that have been used in applications including environmental and industrial catalysts.[50] Both these active doped titanium dioxides and undoped titanium dioxide have been used as photocatalysts. An increased rate in photocatalytic reaction is observed as the redox potential increases and the size decreases. Such additives can be used as a component in self-cleaning paints and plasters. Photocatalytic titanium dioxide can decompose organic substances when it absorbs light. One use has been in self-cleaning windows. Another is the "bathroom that cleans itself", where self-cleaning tiles treated with nanoparticulate titanium dioxide may be found. The titanium dioxide nanoparticles absorb light and microbes on the surface are destroyed. The removal of nitrogen oxides from the atmosphere using photoactive titanium dioxide[51] and removal of contaminants from water have also been reported.[52]

Nano titanium dioxide has also been used in solar cells as the active component for absorption of solar energy. The nanocrystalline titanium dioxide dye-sensitised solar cell was originally developed to overcome the problems experienced by conventional solar cell technology.[53–55]

3.4.2 Zinc Oxide. While titanium dioxide dominates the inorganic UV absorption market, ultrafine zinc oxide is used in similar applications although at smaller volumes. Products are on sale from among others BASF, Nanophase, Umicore and Advanced Nanoproducts. It is claimed that nano zinc oxide results in a more transparent coating than an equivalent coating containing nano titanium dioxide.[56] Doped variants of zinc oxide may also be produced, with Oxonica again exploring the potential for a manganese-doped material.

3.4.3 Aluminium Oxide. Nanoparticulate aluminium oxide has been produced in platelet form and has found use in cosmetics. The benefits are achieved through a uniform platelet morphology that provides superior transparency and soft focus properties.[57]

3.4.4 Silicon Dioxide. When Degussa chemist Harry Kloepfer invented a process to produce an extremely fine silicic acid in 1942, he had no idea that this would mark the first chapter in an extraordinary success story that is still continuing today.[58] Silicic acid, better known today as fumed silica and marketed under the name Aerosil by Degussa since 1943, is now produced in a large number of variants and sold to almost 100 countries worldwide, and other companies including Cabot Corporation also produce and supply their own version of the material. Kloepfer had originally developed the substance as an alternative to carbon blacks as a reinforcing filler for car tyres.

Fumed silica has a chain-like particle morphology. In liquids, the chains bond together via weak hydrogen bonds forming a three-dimensional network, trapping liquid and effectively increasing viscosity. The effect of the fumed silica can be negated by the application of a shear force, *e.g.* by mixing or spraying, allowing the liquid to flow and level out and permitting the escape of entrapped air. However, when the force is removed, the liquid will "thicken up". This property is called thixotropy and products exploiting this characteristic of fumed silica include non-drip paint. When added to powders, fumed silica aids flow and helps prevent caking so the product is also used with other fillers as additives in plastics where effective dispersion is key to performance. Such products include adhesives, coatings, cements and sealants. Fumed silica also finds use in cosmetics, pharmaceuticals, pesticides, inks, batteries and abrasives. The total market for fumed silica is in excess of 1 million tonnes per year.

3.4.5 Iron Oxide. Nano forms of iron oxide have found application in cosmetics and in catalysts, including catalysts for enhanced oxidation of diesel fuel and soot derived from diesel fuel either alone or in combination with cerium oxide. An example of this employs a combination of iron and cerium compounds that are oxidised to the oxides in the combustion chamber of diesel engines and when these oxides interact with soot in the diesel particulate filter the combustion of the soot is catalysed with the result that there is a shorter regeneration time for the filter.[59]

3.4.6 Cerium Oxide. Cerium oxide is a well-known oxidation catalyst and has been used in a variety of forms in a number of products. However, to exploit its catalytic activity most effectively, nanoparticulate cerium oxide has been used successfully as a catalyst for enhancing the combustion of diesel fuel to reduce emissions and reduce fuel consumption. A product called Envirox from Oxonica is based on nanoparticulate cerium oxide and the cerium oxide is delivered to the engine in the diesel fuel at a level of 5 ppm.[60]

3.5 Clays

Naturally occurring complex molecules such as clay can be treated to release nanometre scale platelet structures. These materials, with their ability to align to produce barrier layers, have been used in a number of applications where a gas barrier is required or where reinforcement is required in a single dimension. The essential nanoclay raw material is montmorillonite, a 2-to-1 layered smectite clay mineral with a platelet structure, and is based on magnesium aluminium silicate. Individual platelet thicknesses are just one nanometre, but surface dimensions are generally 300 to more than 600 nanometres, resulting in an unusually high aspect ratio. Naturally occurring montmorillonite is hydrophilic and, since polymers are generally hydrophobic, unmodified nanoclay disperses in polymers with great difficulty. Through clay surface modification, montmorillonite can be made hydrophobic and therefore compatible with conventional polymers.

Compatibilised nanoclays disperse readily in polymers including nylon, polyethylene, polypropylene, PVC and polystyrene. Applications exploit the platelet form of the nanoclay where the platelets align themselves improving barrier properties, increasing modulus and tensile properties and increasing flame retardancy. As an example of what can be achieved, nanocomposites containing nanoclays look attractive for moulded car parts as well as for electrical/electronic parts and appliance components. On the packaging side, nanocomposites can slow transmission of gases and moisture vapour through plastics by creating a "tortuous path" for gas molecules to thread their way among the obstructing platelets. Bottles and food packaging are not the only areas of interest.

Nanocomposites hold commercial benefits for reducing hydrocarbon emissions from hoses, seals and other fuel system components. Flame retardant properties of nanocomposites are of interest on many fronts. Reduced flammability of nanocomposites has been demonstrated for several different thermoplastics including polypropylene and polystyrene. One application that has novelty value is a new tennis ball produced by Wilson. This ball has a nanocomposite coating which it is reported "keeps it bouncing twice as long as a conventional one". This results from the reduction of gas transmission through the wall of the tennis ball.

3.6 Quantum Dots

A quantum dot is a semiconductor nanocrystal whose size is in the range 1–10 nm. The size of these particles results in new quantum phenomena that yield significant benefits. Material properties change dramatically at this scale because quantum effects arise from the confinement of electrons and holes in the material. Size changes other material properties such as the electrical and nonlinear optical properties of a material making them very different from those of the material's bulk form. If a dot is excited, the smaller the dot, the higher the energy and intensity of its emitted light. Hence these very small semiconducting quantum dots provide the potential for use in a number of new applications. The colour of the emitted light depends on the size of the dot: the larger the dot, the redder the light. As the dots become smaller, the emitted light becomes shorter in wavelength yielding emitted blue light.

Quantum dots may be metallic, for example gold, or chalcogenide based, *e.g.* cadmium selenide or sulfide. Given that a rainbow of colours is at least theoretically possible, dependent on the size and chemistry of quantum dots, a number of interesting applications are currently being developed. Light-emitting diodes of different colours have been produced, with white light production also possible using a combination of dots. Multi-colour lasers may be developed based on these particles.[61]

When coated with a suitable chemically active surface layer, quantum dots can be coupled to each other or to different inorganic or organic entities and thus serve as useful optical tags. The use of this characteristic of quantum dots is probably most evident in studies in biology and medicine.[62,63] The photo-luminescence as defined by the combination of the size and chemistry of the quantum dot may be exploited in bioanalytical applications. Previously these applications have used organic dyes. However, the use of quantum dots may allow for high sensitivity multiplexed methods, due to their narrow and intense emission spectra. This is in contrast to organic fluorophores, which suffer from fast photobleaching and broad overlapping emission lines. This limits their application considerably.

To make quantum dots useful for such assays they need to be conjugated to biological molecules, which may then be reacted to an active species in the test. Applications include both *in vitro* and *in vivo* use. Specificity is one of the most critical criteria for measuring particular molecules and the characteristics of quantum dots lend themselves to addressing such problems.

3.7 Surface Enhanced Raman Spectroscopy

An alternative route to achieving the same specificity uses either gold or silver cores at a size of approximately 20 nm surrounded by a marker molecule such as a dye and further surrounded by a polymer or inorganic coating such as silica, which allows conjugation with appropriate biological molecules. This is Surface Enhanced Raman Spectroscopy, or SERS, and the Raman spectrum emitted from this combination in response to light stimulation is unique and offers a similar capability to determine active biological species, but a at a much lower concentration than with quantum dots. Products based on this technology are currently under development by Oxonica.[64]

3.8 Dendrimers

Although linear polymers may be considered to be of nanomeric dimensions, there is one specific group of polymers that is designed to exploit its nanomeric size and characteristics. These are dendrimers and they are large and complex molecules with very well-defined structures. They are almost perfectly monodisperse macromolecules with a regular and highly branched three-dimensional architecture. Dendrimers can act as biologically active carrier molecules in drug delivery, to which can be attached therapeutic agents. They can also be used as scavengers of metal ions, offering the potential for environmental clean-up operations.[65]

A dendrimer is a macromolecule which is characterised by its highly branched three-dimensional structure. The structure is always built up around a central multi-functional core molecule and this extremely regular structure contributes to its near-perfect spherical shape. Due to their size, *c*. 15 nm, and branching architecture with a relatively hollow core surrounded by a compact surface, dendrimer molecules could be utilised for sensing, catalysis or biochemical activity. They may also find application as light-harvesting antennae and as molecular amplifiers.[66] It has also been suggested that when drug molecules are attached to the periphery, the dendrimer can be used as an efficient drug-delivery platform. Studies have demonstrated potential application of dendrimers as gene carriers.[65]

4 Bio Applications

Nanotechnology provides the tools to measure and understand biosystems. Applications of nanotechnology to biotechnology, biomedicine and agriculture include biocompatible implants, manipulation of molecules within cells, biocompatible electronic devices and "smart" controlled release delivery of nutrients.[67–69] Nano-oncology offers promise in cancer treatment with the potential for delivery of anticancer drugs and the localised killing of cancerous and precancerous cells[70] or for more general drug delivery[71] with some potential for drug delivery across the blood–brain barrier.[72] Nanotubes have also been considered for delivery of active species or for separating and collecting active species, but this technology is still in its infancy.[73]

5 Nanocatalysts

Cerium oxide is only one example of a nanocatalyst. Many nanocatalysts derive their activity simply from the large increase in surface area associated with nanoparticles. The global market for nanocatalysts is projected to approach \$5 billion in 2009.[74] Commercially, well-established nanocatalysts such as industrial enzymes, zeolites and transition metal nanocatalysts accounted for about 98% of global sales in 2003. Newer types such as transition metal oxides, metallocenes, asymmetric carbon nanotubes and others are expected to grow significantly through to 2009. The refining/petrochemical sector was the largest user in 2003 with over 38% of the market, followed by chemicals/pharmaceuticals, food processing and environmental remediation.

6 Nanotechnology Reports

6.1 Forbes/Wolfe Nanotech Reports

Forbes/Wolfe produce a monthly newsletter on nanotechnology called Nanotech Report and at the end of each year report on the top 10 nanotech products of the year. In 2004, the products included a nanotechnology-based footwarmer containing a nanoporous aerogel, golf clubs using "titanium fullerene materials" in the head of their new driver, nanosilver-containing wound dressings with improved antibacterial effectiveness, an additive from BASF that improves the hydrophobicity of building materials and silica nanofillers in dental adhesives.[75]

In 2005, the follow-up report on the top 10 nanotech products led with Apple's iPod Nano as the number one product, but whether this product represents nanotechnology or is simply marketing hype was the question to consider.[76] The report concludes that the answer to both parts of the question is a resounding "Yes" in that the nano connection certainly attracted attention, but inside the product there are memory chips that are produced with precision less than 100 nm.

Given the range of cosmetics using nanoparticulate metal oxides primarily for UV protection it is interesting to note a cosmetics product containing fullerene in the list. In this case the fullerene is claimed to have antioxidant properties. Carbon nanotubes have been used as a reinforcing component in a new baseball bat. Silver nanoparticles feature again, this time in socks where enhanced bonding of the 19 nm silver particles to the polyester fibres is claimed to provide enhanced and longer-lasting antimicrobial and antifungal performance. A novel chewing gum having chocolate flavour, which is apparently difficult to achieve, has been produced using "nanoscale crystals" of unknown chemistry to enhance the compatibility of the cocoa butter with the polymers that are used to give the gum elasticity. So-called self-cleaning windows and paint surfaces are also included in the top 10. These are based on photoactive titanium dioxide with the windows gaining a further benefit when it rains, with the hydrophilic film created being washed off leaving a clear surface.

6.2 Woodrow Wilson

The Project on Emerging Nanotechnologies is an initiative by the Woodrow Wilson Center and the Pew Charitable Trusts in 2005. As part of this initiative the Project has launched The Nanotechnology Consumer Products Inventory.

This is the first online inventory of nanotechnology consumer products and contains some 212 manufacturer-identified nanoproducts. The inventory can be accessed online at www.nanoproject.org/consumerproducts and at least some of the products and applications described here are listed in this inventory. Others include reinforced tennis, squash and badminton racquets containing carbon nanotubes, cultured diamonds, non dirtying clothes, razors, automotive and other coatings, cosmetics, microprocessors, golf balls, silver colloids and photographic paper.

7   Future Opportunities

7.1   Nanoroadmap

The Nanoroadmap Project has been co-funded by the European Commission as part of their Framework 6 initiative and has produced a document in late 2005 as a report in four parts, *i.e.* Nanoporous Materials, Nanoparticles/ Nanocomposites, Dendrimers and Thin Film and Coating.[77] The reader is directed to this report for the detail. Nanoparticle applications are considered under the headings power/energy, healthcare/medical, engineering, consumer goods, environmental and electronics, and potential applications are considered through to 2015. Some of these are based on technologies discussed here and include solar cells, fuel cells and automotive catalysts, fungicides, nanoclay/polymer composites, inks, chemical sensors, photocatalysts, optoelectronic devices, biolabelling and detection and new dental composites.

Nanostructures including thin films and coatings are also considered and applications there reflect at least some of the opportunities for nanoparticles in the future such as solar cells and self-cleaning surfaces, but also include superconductivity applications and thin-film transistors.

7.2   SusChem

The European Technology Platform (ETP) for Sustainable Chemistry (SusChem) was initiated jointly by Cefic and EuroBio in 2004 to help foster and focus European research in chemistry, chemical engineering and industrial biotechnology. The SusChem vision foresees a sustainable European chemical industry with enhanced global competitiveness, providing solutions to critical demands and powered by a world-leading innovative drive. SusChem unites a wide variety of stakeholders around this common vision. This process is designed to elicit programme areas that should be funded by the EU as part of its Framework 7 initiative to begin in 2007. Thus needs have been identified and potential programmes are sought to align with those needs.

A recently published document represents the current Strategic Research Agenda of SusChem and the Materials Technology section focuses on six areas of need for the future.[78] These are Energy, ICT, Healthcare, Quality of Life, Transportation and Citizen Protection. Underpinning the product development to address the needs for the future in each of these six areas are new materials with nanoscience seen as the basis for development of the new materials. In general the potential of nanoscience lies in the ability to provide new applications in the fields of catalysis, higher reactivity in synthesis, better biocompatibility and enhanced electrical and mechanical properties. Industry was encouraged to think of nanotechnology as an innovation toolkit that can lead to new materials at the nanoscale which spawn new products and ideas for the market and assist in creating new markets.

7.3 Lux Research Market Forecast

Lux Research is a leading research and advisory firm specialising in the business and economic impact of nanotechnology and related emerging nanotechnologies. They have recently produced a report forecasting that the value of products incorporating nanotechnology will total $2.6 trillion in 2014.[79] They define nanotechnology as a set of tools and processes for manipulating matter that can be applied to virtually any manufactured goods. They consider that the value of basic nanomaterials will be of the order of $13 billion in 2014.

Through 2009, electronics and IT applications are considered likely to dominate as microprocessors and memory chips built using new nanoscale processes come to the market. They envisage that nanotechnology will become commonplace from 2010 onwards as commercial breakthroughs over the next four years are converted into products. Healthcare and life sciences applications will most likely become significant during this period as nano-enabled pharmaceuticals and medical devices come to the market.

8 Nanomaterials Companies

For those interested in seeking out the wide range of companies that are currently producing nanoparticles of many different types, the reader is advised to check the following website: www.nanovip.com/enventory.Materials/index. php. There a brief overview of each of the companies listed is accompanied in many cases by a detailed profile of the company.

9 Future

If the forecast from Lux Research is to be believed, then there will be further very significant growth of the use of nanomaterials and a reliance on nanotechnology over the next ten years and beyond. The major companies that have been active in nanomaterials for many years continue to invest heavily in new products, and in Japan and China there has been a very significant growth in investment in this whole area that will inevitably lead to products that may not even have been considered today. Given the range of products and applications described here and this investment for the future, future applications of nanotechnology will be many and will excite the scientist and consumer alike.

References

1. N. Taniguchi, *Proc. Intl. Conf. Prod. Eng. Tokyo, Part II*, Japan Society of Precision Engineering, 1974.
2. K. Eric Drexler, *Engines of Creation: The Coming Era of Nanotechnology*, 1986.
3. The Royal Society and Royal Academy of Engineering, *Nanoscience and nanotechnologies: opportunities and uncertainties*, 2004.
4. http://www.ipt.arc.nasa.gov/nanotechnology.html, 2006.
5. M.C. Roco, *J. Nanoparticle Research*, 2005, **7**, 707.
6. http://www.nano.gov/html/res/IntlFundingRoco.htm.
7. http://www.nano.gov/html/about/funding.html.
8. *Carbon: Science and Technology*, 2nd edn, 1993, Ed J.-B. Donnet et al., Marcel Dekker.
9. *Conductive Polymers and Plastics in Industrial Applications*, 1999, Ed L. Rupprecht, William Andrew Publishing/Plastics Design Library.
10. http://www.cancer.com/.
11. http://www.coatings.de/radcure/reading/pietschmann.htm.
12. J. Segurola, N.S. Allen, M. Edge, A. Parrondo and I. Roberts, *J. Coat. Technol.*, 1999, **71**, 61.
13. http://www.allbusiness.com/periodicals/article/501514-1.html, 2005.
14. T. Enoki, Y. Kobayashi, N. Kawastsu, Y. Shibayama, B. Prasad, H. Sato, K. Takai and K. Harigaya, 10th Conference on Molecular Nanotechnology, 2002.
15. S. Patchkovskii, J.S. Tse, S.N. Yurchenko, L. Zhechkov, T. Heine and G. Seifert, *Proc. Natl. Acad. Sci. U.S.A.*, 2005, **102**, 10439.
16. http://physicsweb.org/articles/news/9/7/10.
17. H.W. Zhu, C.L. Xu, D.H. Wu, B.Q. Wei, R. Vajtai and P.M. Ajayan, *Science*, 2002, **296**, 884.
18. C. Dekker, *Phys. Today*, 1999, **52**, 22.
19. M.R. Falvo, G.J. Clary, R.M. Taylor, V. Chi, F.P. Brooks, S. Washburn and R. Superfine, *Nature*, 1997, **389**, 582.
20. http://physicsweb.org/articles/world/13/6/8, 2000.
21. R.H. Baughman, A.A. Zakhidov and W.A. de Heer, *Science*, 2002, **297**, 787.
22. T. Yildirim and S. Ciraci, *Phys. Rev. Lett.*, 2005, **94**, 175501.
23. Y. Zhao, Y.-H. Kim, A.C. Dillion, M.J. Heben and S.B. Zhang, *Phys. Rev. Lett.*, 2005, **94**, 155504.
24. http://pubs.acs.org/cen/news/83/i35/8335notw1.html.
25. M. Remskar, *Adv. Mater.*, 2004, **16**, 1497.
26. C.N.R. Rao and M. Nath, *Dalton Trans.*, 2003, 1.
27. G.R. Patzke, F. Krumeich and R. Nesper, *Angew. Chem. Intl. Ed.*, 2002, **41**, 2446.
28. R. Tenne, L. Margulis, M. Genut and G. Hodes, *Nature*, 1992, **360**, 444.
29. J. Cumings and A. Zettl, *Chem. Phys. Lett.*, 2000, **318**, 497.
30. O.K. Varghese, G.K. Mor, C.A. Grimes, M. Paulose and N. Mukherjee, *J. Nanosci. Nanotechnol.*, 2004, **4**, 733.

31. C.K. Varghese, M. Paulose, K. Shankar, G.K. Mor and C.A. Grimes, *J. Nanosci. Nanotechnol.*, 2005, **5**, 1158.
32. http://www.nanoscale.com/markets.asp, 2006.
33. W.X. Zhang, *J. Nanopart. Res.*, 2003, **5**, 323.
34. V.F. Puntes, K.M. Krishnan and P. Alivisatos, *Appl. Phys. Lett.*, 2001 **78**, 2187.
35. V.F. Puntes, K.M. Krishnan and A.P. Alivisatos, *Science*, 2001, **291**, 2115.
36. C. Petit, A. Taleb and M.P. Pileni, *J. Phys, Chem. B*, 1999, **103**, 1805.
37. S. Sun and C.B. Murray, *J. Appl. Phys.*, 1999, **85**, 4325.
38. Q.A. Pankhurst, J. Connolly, S.K. Jones and J. Dobson, *J. Phys. D: Appl. Phys.*, 2003, **36**, R167.
39. J.M. Nam, C.S. Thaxton and C.A. Mirkin, *Science*, 2003, **301**, 1884.
40. I. Sondi and B. Salopek-Sondi, *J. Colloid Interface Sci.*, 2004, **275**, 177.
41. http://www.nanopchem.com/, 2006.
42. C. Chen, L. Wang, G.H. Jiang and H.J. Yu, *Rev. Adv. Mater. Sci.*, 2006, **11**, 1.
43. C.Z. Li, H.X. He, A. Bogozi, J.S. Bunch and N.J. Tao, *Appl. Phys. Lett.*, 2000, **76**, 1333.
44. P.N. Bartlett and S. Guerin, *Anal. Chem.*, 2003, **75**, 126.
45. A.G. Tkachenko, H. Xie, D. Coleman, W. Glomm, J. Ryan, M.F. Anderson, S. Franzen and D.L. Feldheim, *J. Am. Chem. Soc.*, 2003, **125**, 4700.
46. L.R. Hirsch, R.J. Stafford, J.A. Bankson, S.R. Sershen, B. Rivera, R.E. Price, J.D. Hazle, N.J. Halas and J.L. West, *Proc. Natl. Acad. Sci. U.S.A.*, 2003, **100**, 13549.
47. G. Wakefield, J. Stott and J. Hock, *SOFW*, 2005, **131**, 46.
48. G. Wakefield, J. Stott and Forbes N., *Soap, Perfumery and Cosmetics*, April 2005, 41.
49. G. Wakefield, S. Lipscomb, E. Holland and J. Knowland, *Photochem. Photobiol. S.*, 2004, **3**, 648.
50. http://www.millenniumchem.com.
51. F.L. Toma, G. Bertand, D. Klein and C. Coddet, *Environ. Chem. Lett.*, 2004, **2**, 117.
52. R.W. Matthews, *Pure Appl. Chem.*, 1992, **64**, 1285.
53. Q. Dai and J. Rabani, *J. Photochem. Photobiol., A*, 2002, **148**, 17.
54. A.K. Jana and B.B. Bhowmik, *J. Photochem Photobiol., A*, 1999, **122**, 53.
55. B. Oregan and M. Gratzel, *Nature*, 1991, **353**, 737.
56. http://en.wikipedia.org/wiki/Zinc_oxide.
57. Advanced Nanotechnologies Ltd, http://www.advancednanotechnology.com/alusion.php, 2006.
58. http://www.degussa.com/en/home.html.
59. DieselNet Technology Guide.
60. B. Park, R. Scattergood, C. Harris, G. Goddard and S. Samuel, *Additive 2005: Optimising Automotive Power Trains*, Dublin, 2005.
61. V.I. Klimov, *Los Alamos Science*, 2003, **28**, 214.
62. A.P. Alivisatos, W. Gu and C. Larabell, *Annu. Rev. Biomed. Eng*, 2005 **7**, 55.

63. R.E. Bailey, A.M. Smith and S.M. Nie, *Physica E*, 2004, **25**, 1.
64. Faraday Discussion 132: Surface Enhanced Raman Spectroscopy, 2006, 132, 1.
65. M. Guillot, S. Eisler, K. Weller, H.P. Merkle, J.L. Gallani and F. Diederich, *Org. Biomol. Chem.*, 2006, **4**, 766.
66. O. Flomenbom, R.J. Amir, D. Shabat and J. Klafter, *J. Lumin.*, 2005 **111**, 315.
67. D.F. Emerich and C.G. Thanos, *Expert Opin. Biol. Ther.*, 2003, **3**, 655.
68. W.J. Parak, T. Pellegrino and C. Plank, *Nanotechnology*, 2005, **16**, R9.
69. M.C. Roco, *Curr. Opin. Biotechnol.*, 2003, **14**, 337.
70. B. Panchapakesan, *Oncol. Issues*, 2005, 20.
71. J. Panyam and V. Labhasetwar, *Adv. Drug Deliver. Rev.*, 2003, **55**, 29.
72. P.R. Lockman, R.J. Mumper, M.A. Khan and D.D. Allen, *Drug Dev. Ind. Pharm.*, 2002, **28**, 1.
73. P. Kohli and C.R. Martin, *Curr. Pharm. Biotechnol.*, 2005, **6**, 35.
74. Business Communications Company, Nanocatalysts, 2004.
75. Forbes/Wolfe Nanotech Report, 2004.
76. Forbes/Wolfe Nanotech Report, 2005.
77. C. Escolano, J. Perez and L. Bax, Nanoroadmap Project-Roadmap reports: Materials, 2005.
78. The European Technology Platform for Sustainable Chemistry (SusChem), 2005.
79. Lux Research Report.

Nanoparticles in the Aquatic and Terrestrial Environments

JAMIE LEAD

1 Introduction

Engineered nanoparticles are being produced in increasing amounts and are being discharged to the aquatic and terrestrial environments in considerable volume. Both production and discharge are likely to increase substantially in the near future. For overviews of current and projected investment and production a number of reviews are available.[1,2] These nanostructures are of concern as they will interact with aquatic and terrestrial systems in largely unknown ways, are potentially deleterious to ecological health, may be vectors of pollution and because their transport and ecotoxicology are essentially unknown. This chapter will review the available literature, and address the potentially relevant issues in relation to aquatic and terrestrial systems. It is hoped that at this early stage this chapter will help to clarify the areas which need to be addressed, but few definitive answers can be provided because the knowledge base is so scanty.

Engineered nanoparticles, nanotubes and other structures may be defined as anthropogenically produced material between 1 nm and 100 nm in size,[1] although no formal and accepted definition is available as yet. This area has received increasing attention and concern in recent years on an international scale.[3] There are three reasons for this interest and concern in relation to the natural aquatic and terrestrial environments:

(i) These materials are being produced in ever greater amounts in both research and industrial processes.
(ii) These nanostructures may behave in significantly different ways from larger bulk material, even where structures are chemically similar.
(iii) Their impacts on human and ecological health are largely unknown, but are potentially severe.

Issues in Environmental Science and Technology, No. 24
Nanotechnology: Consequences for Human Health and the Environment

The reports mentioned above discuss the first point *i.e.* overall amounts of nanostructures produced and intentional and accidental discharges to aquatic environments. Discharges may potentially be from point sources such as specific industrial waste streams and injection into contaminated land as a remediation procedure and from diffuse sources such as cosmetics and sunscreens being washed off individuals. Focus has recently been on the "free" form of the nanoparticles, but as discussed later, discharge to the environment of "fixed" forms of nanoparticles may also be important. On the second point, differences in the behaviour are generally attributed to both surface area and quantum effects. For instance, the rate of oxidation of Mn by nanoscale haematite was shown to be dependent on size,[4] with effects due to geometrical and electronic changes with size. Effects were partly due to specific surface area differences, but once normalised the smaller size particles (*ca.* 7 nm) increased the rate of oxidation by more than an order of magnitude in relation to the larger particles (*ca.* 35–40 nm). This chapter will attempt to address point iii (at least in principle) in relation to natural waters.

2   Overview of Current Knowledge

Direct knowledge of the fate, behaviour and ecotoxicology in natural aquatic systems is extremely limited, but there are extensive "review" and discussion documents both in the peer-reviewed literature and other sources. These appear to be based on spectacularly small amounts of real data. Concern is therefore warranted since (a) we do not always understand natural aquatic systems with great depth and confidence, (b) we have almost no direct knowledge of engineered nanoparticles, even in artificial laboratory settings, and (c) there is little or no specific regulatory framework for the control of nanoparticle production and disposal to the aquatic environment. All these issues are in the process of being addressed but clearly time is required to make progress in all areas. Meanwhile, funding and developments in nanotechnology continue to accelerate while relatively limited attention is given to the possible human and environmental health effects of nanotechnology. A short but comprehensive literature survey of relevant data is therefore presented. However, the paucity of data becomes apparent when we also consider the complexity and variability of natural systems in terms of their chemistry, hydrology and ecology and the range and types of nanoparticle structures which are dependent on size, chemistry and three-dimensional architecture. Only a few peer-reviewed papers have been produced and these cover a narrow range of essentially artificial conditions.

Toxicity data on engineered nanoparticles in humans[5,6] and other mammals[7] are becoming more available, although they are still sparse. A number of these nanoparticles such as organic fullerenes and carbon nanotubes and inorganics such as Ag particles have been shown to be harmful to organisms. In aquatic systems, direct evidence on the effect of fullerenes in both freshwater fish and bacteria is available. Fullerenes are sparingly water-soluble but form aggregates between 5 and 1000 nm in size in waters, dependent upon prior treatment.

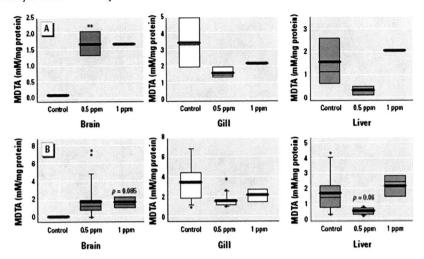

Figure 1 Lipid peroxidation of brain, gill and liver tissue in largemouth bass on exposure to aqueous fullerene suspensions. (A) are averages for all fish and (B) data for individual fish. Thick black lines: mean values; thinner lines: median values; boxes represent 25th and 75th percentiles and error bars are ranges. Taken from reference 8.

The effects of polydisperse C_{60} aggregates (30–100 nm) on juvenile largemouth bass have been investigated.[8] Significant effects on brain lipid oxidation at 0.5–1.0 part per million (ppm) concentrations of the fullerenes were noted, as shown in Figure 1. No evidence of oxidation was found in lipid oxidation of gill or liver tissues nor was there evidence of protein oxidation of any tissues, although possible impacts on glutathione levels were noted. Qualitative indications on bacterial behaviour were also noted, primarily from the clarification of both water and container glass. This agrees with data on impacts of aqueous fullerene aggregates on both Gram-negative (*E. coli*) and Gram-positive (*B. subtilis*)[9] bacterial species as is shown in Figure 2. Fullerenes in this study were between 20 and 180 nm, with a variety of conformations (spherical, rectangular and some triangular). At concentrations of 4 mg L^{-1}, significant reduction in bacterial growth was observed. Similar effects were not observed with fullerenols (OH substituted fullerenes). Other information on significant toxic effects on bacteria relevant to natural aquatic systems has confirmed these data[10] and suggested possible mechanisms, including the disruption of the electron transport chain, physical disruption of cell membranes and production of reactive oxygen species. Initial results are ambiguous[9] and the mechanisms of antimicrobial effects require elucidation. Studies on powdered fullerenes have shown no such impact.[10] While other toxicology data exist in the literature, few if any are relevant to natural aquatic systems.

Aggregation behaviour and size of nanoparticles has been noted in the literature and this will have implications for the fate and behaviour of these

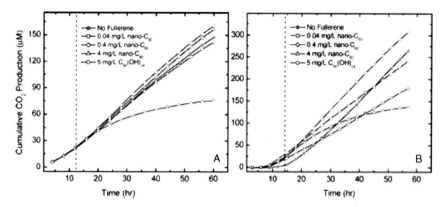

Figure 2 Response of *E. coli* (A) and *B. subtilis* (B) to aqueous suspensions of fullerenes and fullerenols as measured by carbon dioxide production. Taken from reference 9.

nanoparticles in the environment. Several authors have found zeta potentials of roughly –10–50 mV for fullerenes at environmentally relevant pH and low ionic strength values.[9,11] Inorganic phases are generally also negatively charged, except for iron oxide materials, which tend to be positively charged under most environmentally relevant conditions.

Fullerenes may exist in molecular form in organic solvents, but tend to aggregate into small clusters in water and at larger ionic strengths aggregate further to sizes outside the nanoparticle range. Zeta potentials of fullerenes have been shown to be dependent on ionic strength, with decreasing values with increasing ionic strength, indicating the possibility of significant aggregation in estuarine and marine conditions. An example of particle size distributions from dynamic light scattering measurements in response to ionic strength is shown in Figure 3. A number of possible mechanisms for the formation of fullerene surface charge have been postulated, but its origin remains unclear.

Iron oxide nanoparticles are stable at low pH values.[12] However, increase in pH or ionic strength induces significant aggregation and losses from water, indicating that the particles are again stabilised by charge repulsion mechanisms. It is also well established that natural organic macromolecules (NOM) will interact with iron nanoparticles forming surface films several nanometres thick[13–15] with changes in the behaviour of the iron in the environment. Force–distance curves for fixed iron oxide nanoparticles in the presence and absence of natural organic macromolecules derived from atomic force microscopy (AFM) are depicted in Figure 4. Clear changes are observed and electrostatic, steric and other binding mechanisms have been deduced. The organic layers have been estimated as being about 3–5 nm thick,[15,16] although films up to about 100 nm have been observed on macroscopic surfaces.[17] Changes observed on nanoparticle forms by AFM agree well with electrophoretic data,[18,19] which show that electrophoretic mobilities are controlled by the sorbed NOM.

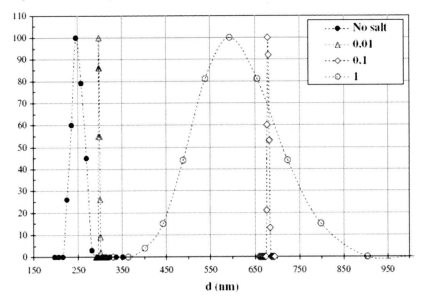

Figure 3 Size distributions of fullerene aggregates in water as a function of ionic strength. Taken from reference 11.

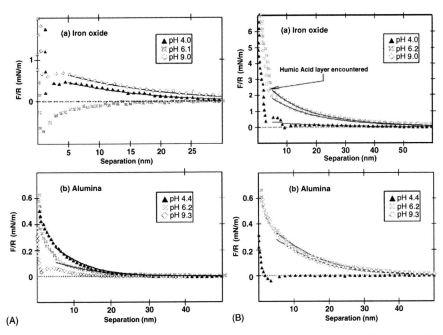

Figure 4 Force–distance curves from AFM showing the approach curves of a 5 mm Si bead to either fixed iron oxide or alumina nanoparticles in the absence (A) or presence (B) of natural organic macromolecules. Taken from reference 14.

Movement of nanoparticles in porous media (soils, groundwaters and laboratory analogues, *e.g.* columns of glass beads) have been studied somewhat more frequently in relation to relevant transportation issues,[11,20,21] especially in relation to the movement of Fe nanoparticles used in the remediation of contaminated land.[22-24] The results generally indicate that transportation is often rapid and with minimal capture by the solid phase, due to repulsive forces between nanoparticles and low collision rates. Nevertheless, this is highly dependent on the particle, column and solution conditions. For instance, unsupported Fe nanoparticles aggregate rapidly and do not travel well, while Fe coated in synthetic polyelectrolytes are sufficiently charge stabilised to prevent aggregation and increase transport.[24] Additionally, Fe nanoparticles did not aggregate rapidly in clay-rich soils due to the formation of anionic surface films from the clay which prevented rapid aggregation. In these cases, the clay and synthetic polyelectrolytes acted in a similar manner to the NOM mentioned in the previous paragraph.

Although these data are generally collected from relatively simple solutions, conclusions can be drawn about the environmental behaviour of such materials. For instance, it might be assumed that aggregation might reduce transport and limit any negative effects of nanoparticles.[11] Even where this occurs, it would be somewhat simplistic since, in surface waters, aggregation would be followed by sedimentation, loss from the water column and build-up of concentrations in the sediments of marine and fresh waters. Similar processes have been observed with metal pollution where naturally occurring nanoparticles become involved in a process ("colloidal pumping"), which results in uptake of metals to nanoparticles, followed by aggregation and sedimentation and build-up of metal concentrations in the sediments.[25] Concentration profiles of metals in sediments dated with *e.g.* Pb-210 dating have allowed trends such as industrialisation to be followed. If similar processes occurred with engineered nanoparticles, localised, rather than more widespread, effects of the nanoparticles may be observed. However, sediment dwelling organisms, filter feeders and other organisms may be those most at risk from uptake and any potentially deleterious effects of nanoparticles, with possible environmental consequences. Since these organisms include shellfish, there is a significant potential exposure pathway back to humans. However, the available data and *a priori* considerations of likely environmental behaviour indicate that nanoparticles should have considerable mobility. For example, "colloid-facilitated" transport[26,27] may increase the transport of engineered nanoparticles in groundwaters, again with potential routes back to human uptake through contamination of supply waters used for potable water.

Zerovalent iron nanoparticles and other materials have been used extensively in the remediation of contaminated waters and land.[28-31] Figure 5 shows a loss of TCE over time in a large field study, while Table 1 shows the possible contaminants which might be remediated by nanoscale iron. Their success in reducing levels of chlorinated organic chemicals in particular is a noteworthy success story and has led to their increased use on a large scale, especially in the USA. Nevertheless, adverse effects have been noted. Removal of oxygen and

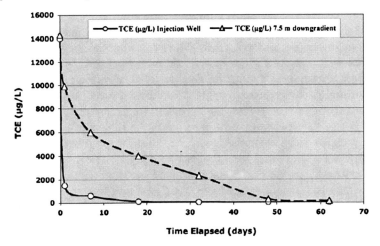

Figure 5 Reduction of trichloroethene concentrations over time after application of zerovalent Fe nanoparticles during an in-field experiment. Taken from reference 28.

Table 1 Common environmental pollutants which can be degraded by zerovalent iron nanoparticles. Taken from reference 28.

Chlorinated methanes	Trihalomethanes
Carbon tetrachloride (CCl_4)	Bromoform ($CHBr_3$)
Chloroform ($CHCl_3$)	Dibromochloromethane ($CHBr_2Cl$)
Dichloromethane (CH_2Cl_2)	Dichlorobromomethane ($CHBrCl_2$)
Chloromethane (CH_3Cl)	Clorinated ethenes
Chlorinated benzenes	Tetrachloroethene (C_2Cl_4)
Hexachlorobenzene (C_6Cl_6)	Trichloroethene (C_2HCl_3)
Pentachlorobenzene (C_6HCl_5)	*cis*-Dichloroethene ($C_2H_2Cl_2$)
Tetrachlorobenzene ($C_6H_2Cl_4$)	*trans*-Dichloroethene ($C_2H_2Cl_2$)
Trichlorobenzene ($C_6H_3Cl_3$)	1,1-Dichloroethene ($C_2H_2Cl_2$)
Dichlorobenzene ($C_6H_4Cl_2$)	Vinyl chloride (C_2H_3Cl)
Chlorobenzene (C_6H_5Cl)	Other polychlorinated hydrocarbons
Pesticides	PCBs
DDT ($C_{14}H_9Cl_5$)	Dioxins
Lindane ($C_6H_6Cl_6$)	Pentachlorophenol (C_6HCl_5O)
Organic dyes	Other organic contaminants
Orange II ($C_{16}H_{11}N_2NaO_4S$ ($C_4H_{10}N_2$))	N-nitrosodimethylamine (NDMA)
Chrysoidine ($C_{12}H_{13}ClN_4$)	TNT ($C_7H_5H_3O_6$)
Tropaeolin O ($C_{12}H_9H_2NaO_5S$)	
Acid Orange	
Acid Red	

creation of anaerobic zones (oddly cited as an environmental benefit by workers in the nanoparticle field[28]) has been demonstrated but most studies do not investigate further possible side-effects. Indeed, in already heavily contaminated land and water, it might be thought that possible deleterious effects may be overlooked since contamination may already be quite severe. Nevertheless, these must be considered since contaminated land often contains biota of intrinsic importance which flourishes only in these contaminated areas because of factors such as lack of biological competition. This has already been shown in acid mine drainage waters.[32] Caution may be warranted before assuming contaminated areas must always need to be remediated. Further, zerovalent Fe would be expected to be extremely reactive but this may not be the case. Stabilisation by anionic surface films, whether natural or synthetic, has been demonstrated.[24] In addition, a possible stabilisation mechanism exists which has recently been postulated for explaining the occurrence of naturally produced nanoscale sulfide in oxic waters, where it would not be expected.[33] Possible mechanisms of stabilisation by natural organic matter such as humic substances have been suggested.[34] Similar mechanisms may stabilise nanoscale iron. If this is the case, transport over long distances is entirely feasible, possibly into pristine environments. The evidence base simply does not exist to predict confidently whether such mechanisms may be important.

Indirectly, interactions between gold nanoparticles and humic substances have been demonstrated.[35] In addition, the research group of the author has produced evidence that there are extensive interactions between natural nanoparticles such as humic substances and polysaccharides with engineered nanoparticles such as fullerenols, nanotubes, gold and iron oxide (manuscripts in preparation).

Clearly, from the brevity of this fairly extensive literature search, it is obvious that few data exist in this area and the data that do exist are taken from largely artificial situations which bear little resemblance to natural conditions. However, consideration of our understanding of natural aquatic systems and processes will give us some insight into possibly important processes.

3   Fate and Behaviour in Natural Aquatic Systems

When considering the fate, behaviour and ecotoxicology of engineered nanoparticles, it is perhaps most obvious to consider naturally occurring nanoparticles and this section will do this. Extensive reviews and discussion pieces in this area are now available[36–39] and so this section focuses on only the main factors of relevance to engineered nanoparticles. These areas are: structural determination and analysis, physical–chemical interaction with pollutants, pathogens and nutrients, impact on uptake of pollutants to organisms, impact on pollutant transport. Although similarities exist, differences also exist between natural and engineered nanoparticles. For instance, natural nanoparticles may be taken up by organisms but are studied primarily for the impacts on pollutant bioavailability, rather than direct toxic effects as with engineered nanoparticles. In addition, the two types of materials may interact, substantially affecting the fate and behaviour of engineered nanoparticles.

3.1 Natural and Engineered Nanoparticle Interactions

Natural solid phase material exists in a wide range of chemical phases and size ranges and combined in complex mixtures. Figures 6 and 7 show examples of these types of phases. Clearly, natural waters cannot be considered as pure

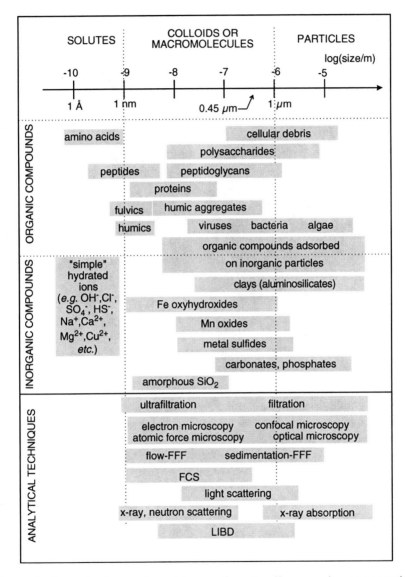

Figure 6 Important organic and inorganic naturally occurring nanoparticles and larger solid phases and important separation and analytical techniques used to characterise them. Taken from reference 37.

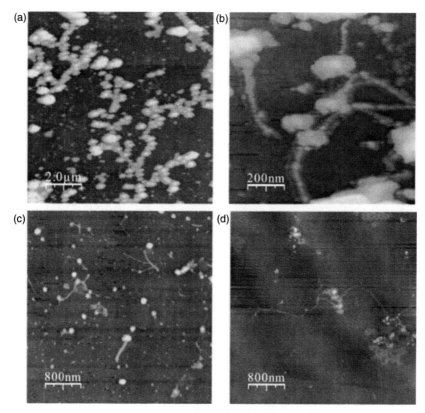

Figure 7   AFM micrographs showing naturally occurring nanoparticles taken
from an urban freshwater. (a) Sample taken from main river; (b)
enlarged view of the fibrils present in image a; (c) and (d) from
tributaries. Taken from reference 17.

waters or simple inorganic solutions, but contain a huge variety of complex,
heterogeneous and polydisperse material from weathering, microbiological
growth and other processes. These suspended forms have the potential to
interact with engineered nanoparticles and substantially alter their behaviour.
To my knowledge, only one published paper[35] deals tangentially with this area
showing that nanoparticle architecture is more easily controlled in the presence
of a natural nanoparticle (fulvic acid). Although with a different focus, this
does show that there must be physico-chemical interactions between natural
and engineered nanoparticles. Indeed, as stated earlier, in the chapter author's
laboratories we have demonstrated significant interactions between several
types of natural and engineered nanoparticles and subsequent effects on
particle size, aggregation and sedimentation (manuscripts in preparation).
For instance, single walled carbon nanotubes in low ionic strength solutions
and filtered (0.1 μm) freshwaters are stable separately over periods of days
alone but when mixed substantial aggregation and sedimentation occurs.

Measurement was by visual inspection, capillary electrophoresis, dynamic light scattering and UV-Vis spectroscopy. No peer-reviewed literature is currently available but this area needs substantial further investigation. The nature of the interactions is likely to be extremely complex due to the different types of natural and engineered nanoparticles and the variability in solution conditions in the environment. These interactions will need to be borne in mind when discussing other processes in the natural aquatic environment, such as transport and the ecotoxicology of engineered nanoparticles.

3.2 Structural Determination and Analysis

This area needs to be commented on only a little, as methods such as transmission electron microscopy are common to the fields of both engineered nanoparticles and aquatic chemistry. Nevertheless, the nature of environmental aquatic samples needs to be considered. In particular, naturally occurring aquatic nanoparticles are hydrated, rather than being dry or present in organic solvents. This clearly has implications for analysis when nanoparticles might have different characteristics in organic solvents and most EM techniques use ultra-high vacuum with consequent dehydration. Choice of appropriate preparation, fractionation and analysis methods will clearly be essential to elucidate the behaviour of engineered nanoparticles in natural waters. Techniques useful in either nanoscience or in aquatic science may not necessarily be applied without modifications, which need to be carefully evaluated.

3.3 Interactions with Pollutants, Pathogens and Nutrients

Research with naturally occurring nanoparticles has indicated that they should be the key chemical species interacting with pollutants and pathogens in natural waters and this is mainly explained by increased specific surface areas at smaller sizes due to geometric considerations.[40] Although little evidence has been published and some is contradictory,[41] evidence of the importance of natural nanoparticles has recently been published in the case of trace elements.[42] We might expect that engineered nanoparticles would also be able to bind large quantities of pollutant material and interact strongly with pathogenic viruses and bacteria. While this may be the case, modifications may occur due to natural nanoparticles. For instance, alteration of the surface charge and chemical reactivity of engineered nanoparticles may occur. In addition, the ubiquity and high concentration of natural material may simply swamp the potential effects of engineered nanoparticles as sorbents of trace pollutants. As we have seen,[4] nanoscale material clearly affects the rates of chemical behaviour in the natural environment.

3.4 Effects on Pollutant and Pathogen Fate and Behaviour

Once bound to engineered nanoparticles, the transport and bioeffects caused by the pollutants are likely to be modified. The pollutant or pathogen will behave as the complex formed with the nanoparticle. This has already been discussed in terms of aggregation and sedimentation in freshwater, affecting transport, and in porous media such as soils and groundwater, "colloid-facilitated transport" may be relevant. In simple terms, the engineered nanoparticle will facilitate transport by providing a binding phase for the pollutant (for instance) by preventing sorption to the immobile solid phase while the pollutant–nanoparticle complex is sufficiently small to be transported through pores in the porous media while being minimally retained. Section 2 has indicated that this is the case and interactions with natural nanoparticles will potentially increase transport rates and overall distances substantially. Transportation rates and distances will be enhanced. Again this process will be influenced by naturally occurring nanoparticles.

Bioavailability of trace elements has also been studied extensively and currently used models such as the free ion activity model (FIAM)[43] and the biotic ligand model (BLM)[44,45] indicate the likely inhibitory effect of binding to nanoparticles and colloids. Nevertheless, considering physical chemical aspects only, these models are simplifications of the real system with severe limitations, many unproved assumptions and they do not fit all available data.[45] In addition, biological processes such as phagocytosis and direct ingestion need to be considered. When considering these ideas in relation to engineered nanoparticles, it may be of interest to consider bound trace pollutants, where these models have direct relationships to the "natural" systems. However, we also need to consider the uptake and toxicity of the engineered nanoparticles directly. In this case, the models are of more limited use, but certainly uptake and depuration kinetics, biochemical and physiological uptake mechanisms as well as hydrological and chemical factors need to be considered. For instance the higher observed toxicity of fullerenes compared to fullerenols in bacteria[11] may be due to the higher solubility of the fullerenes in the cell wall and membrane of the bacteria.

4   Issues to be Addressed

In the absence of much quantitative evidence but the presence of a great deal of speculation, this chapter is inevitably somewhat sketchy. Authoritative answers to the questions raised in this chapter will require a good deal of further work. Some areas which require further consideration are summarised below, building on the previous sections.

4.1   Sources and Sinks of Nanoparticles

Sources are relatively easy to discuss and quantify, at least in principle, since all sources are by definition anthropogenic. Nevertheless, due to the rapid growth and probable changing nature of the industry and other factors, there are likely to be difficulties in practice. Inventories of likely discharges of different types of material need to be kept and regularly updated and these discharges minimised. Sinks are a far harder question to answer, given the probable uncertainties in sinks and our partial knowledge of aquatic systems. Speculation is easy but rigorously collected and interpreted data are lacking at present.

4.2 Free and Fixed Engineered Nanoparticles

Much of the recent speculation on the effects of engineered nanoparticles has focused on the free form, *i.e.* those not physically or chemically bound to larger structures. This has a certain logic given the relative mobility and ease of uptake into organisms of the free form. Nevertheless, the fixed form needs to be considered in aquatic systems because (a) the fixed form must have similar activity to the free form (or else it would not be useful commercially), (b) the fixed form may be colonised by microorganisms and exhibit biological effects and (c) the fixed form will be subject to weathering and/or waste disposal procedures which will convert it to the free form.

4.3 Nanoparticle Interactions with Naturally Occurring Material

This area has been discussed in some depth but it is likely that all aquatic environmental processes will be influenced by this interaction. Urgent further work is required.

4.4 Nanoparticles as Pollutants

Many engineered nanoparticles have been shown to be toxic or potentially so, such as fullerenes, nanotubes, Ag and other inorganic forms. Considerably more work is required given the paucity of our knowledge discussed earlier. Data on toxicity and sub-lethal effects, *e.g.* persistence, reduction in breeding success, *etc.*, need to be produced under realistic conditions, but almost any data are to be welcomed in this area. Specific choice of organism types needs to be made. Investigation of microbial biofilms, due to their ubiquity and importance in biogeochemical cycling, and fish, due to their impact on human health and for economic reasons, are suggested as two important starting points.

4.5 Transport of Nanoparticles

Transportation issues in porous media (soils *etc.*) and surface waters have already been discussed. There is clearly further need for data and appropriate interpretative models on transportation rates and mechanisms. There is also a need for some means of tracking nanoparticles as they move in either lab or field scale. In simple systems (laboratory-scale cores), this can be performed relatively easily using light scattering or other spectroscopic methods. In more complex real systems, there is more difficulty. A possible means of overcoming this problem may be to customise nanoparticles by labelling with stable or radio-isotopes and detecting changes in the signal by inductively coupled plasma mass spectroscopy (ICP-MS) or γ or α counting. Secondly, physicochemical separation by means such as acid digestion, solvent extraction or chromatography may be required prior to analysis. Further analytical validation may be required.

4.6   Nanoparticles as Vectors of Pollution

As discussed, nanoparticles may bind pollutants and alter their behaviour. There are two prerequisites for understanding this process. Firstly, understanding the transport of engineered nanoparticles is required, as discussed. Secondly, the strength of binding (dependent on the strength and number of binding sites *i.e.* binding site density, specific surface area and equilibrium association constants) needs to be assessed under relevant conditions. This work has not been performed, but proven methodological approaches have been applied in other areas and need to be applied here.

5   Conclusions

Further work required is the inevitable conclusion of any researcher in any field and this is usually justified, although useful conclusions can often be drawn as to the way ahead. In the case of the fate, behaviour and ecotoxicology of engineered nanoparticles in aquatic systems, the need for further work is the primary conclusion to be drawn. Secondarily, it is clear that cause for concern is justified and that precautionary measures should be in place to minimise exposure of engineered nanoparticles to the aquatic environment. There is a great deal that can be learnt and employed from the fields of nanotechnology and from aquatic fields but aquatic nanoscience is essentially a new field awaiting our investigation. The next 10 years promise to be a fruitful time in increasing our knowledge. It is to be hoped that the advances in nanotechnology are not deployed so quickly that they outstrip our capacity to understand and offset the potential problems which lie ahead.

References

1. The Royal Society and Royal Academy of Engineering, *Nanosciences and nanotechnologies; opportunities and uncertainties*, The Royal Society, London, 2004, http://www.nanotec.org.uk/finalReport.htm.
2. M.C. Roco, *J. Nanopart. Res.*, 2005, **7**, 707–712.
3. Scientific Committee on Emerging and Newly Identified Health Risks, The appropriateness of existing methodologies to assess the potential risks associated with engineered and adventitious products of nanotechnologies, European Commission, 2005, http://europa.eu.int/comm/health/ph_risk/committees/04_scenhir/scenhir_cons_01_en.htm.

4. A.S. Madden and M.F. Hochella, *Geochim. Cosmochim. Ac.*, 2005, **69**, 389–398.
5. G. Jia, H. Wang, L. Yan, X. Wang, R. Pei, T. Yan, Y. Zhao and X. Guo, *Environ. Sci. Technol.*, 2005, **39**, 1378–1383.
6. A.A. Shvedova, V. Castranova, E.R. Kisin, D. Schwegler-Berry, A.R. Murray, V.Z. Gandelsman, A. Maynard and P. Baron, *J. Environ. Health Toxicol. A*, 2003, **66**, 1909–1926.
7. E. Bermudez, J.B. Magnum, B.A. Wong, B. Asgharian, P.M. Hext, D.B. Warheit and J.I. Everitt, *Toxicol. Sci.*, 2004, **77**, 347–357.
8. E. Oberdörster, *Environ. Health Persp.*, 2004, **112**, 1058–1062.
9. J.D. Fortner, D.Y. Lyon, C.M. Sayes, A.M. Boyd, J.C. Falkner, E.M. Hotze, L.B. Alemany, Y.J. Tao, W. Guo, K.D. Ausman, V.L. Colvin and J.B. Hughes, *Environ. Sci. Technol.*, 2005, **39**, 4307–4316.
10. D.Y. Lyon, L.K. Adams, J.C. Faulkner and P.J.J. Alvarez, *Environ. Sci. Technol.*, 2006, **10**, 4360–4366.
11. J. Brant, H. Lecaotnet and M.R. Wiessner, *J. Nanopart. Res.*, 2005 **7**, 533–545.
12. K. Kendall and M.R. Kossova, *J. Adhesion*, 2005, **81**, 1017–1030.
13. L.M. Mosley, K.A. Hunter and W.A. Ducker, *Environ. Sci. Technol.*, 2003, **37**, 3303–3308.
14. S. Sander, L.M. Mosley and K.A. Hunter, *Environ. Sci. Technol.*, 2004, **38**, 4791–4796.
15. S. Assemi, P.G. Hartley, P.J. Scales and R. Beckett, *Colloid. Surface. A*, 2004, **248**, 17–23.
16. F.J. Doucet, L. Maguire and J.R. Lead, *Anal. Chim. Acta*, 2004, **522**, 59–71.
17. J.R. Lead, D. Muirhead and C.T. Gibson, *Environ. Sci. Technol.*, 2005, **39**, 6930–6936.
18. K.A. Hunter and P.S. Liss, *Limnol. Oceanogr.*, 1982, **27**, 322–335.
19. E. Tipping, C. Woof and D. Cooke, *Geochim. Cosmochim. Ac.*, 1981, **45**, 1411–1414,1419.
20. H.F. Lecoanet and M.R. Wiesner, *Environ. Sci. Technol.*, 2004, **38**, 4377.
21. P. Weronski, J.Y. Walz and M. Elimlech, *J. Colloid Interf. Sci.*, 2003, **262**, 372–383.
22. S.M. Ponder, J.G. Darab, J. Bucher, D. Caulder, I. Craig, L. David, N. Edelstein, W. Lukens, H. Nitsche, L. Rao, D.K. Shuh and T. Mallouk, *Chem. Mater.*, 2001, **13**, 479–486.
23. B. Schrick, J.L. Blough, D. Jones and T. Mallouk, *Chem. Mater.*, 2002, **14**, 5140–5147.
24. B. Schrick, B.W. Hydutsky, J.L. Bough and T.E. Mallouk, *Chem. Mater.*, 2004, **16**, 2187–2193.
25. B.D. Honeyman and P.H. Santschi, in *Environmental Particles*, ed. J. Buffle and H.P. van Leeuwen, Lewis Publishers, Boca Raton, 1992, vol. 1, p. 554.
26. P.H. Santschi, K.A. Roberts and L. Guo, *Environ. Sci. Technol.*, 2002, **36**, 3711–3719.
27. G. Chen, M. Flury, J.B. Harsh and P.C. Lichtner, *Environ. Sci. Technol.*, 2005, **39**, 3435–3442.

28. W.-S. Zhang, *J. Nanopart. Res.*, 2003, **5**, 323–332.
29. I. Dror, D. Baram and B. Berkowitz, *Environ. Sci. Technol.*, 2005, **39**, 1283–1290.
30. S.H. Joo, A.J. Feitz, D.L. Sedlak and T.D. Waite, *Environ. Sci. Technol.*, 2005, **39**, 1263–1268.
31. S.H. Joo, A.J. Feitz and A.D. White, *Environ. Sci. Technol.*, 2004, **38**, 2242–2247.
32. L.C. Batty, *Mine water and the Environment*, 2005, **24**, 101–103.
33. T.F. Rozan, G. Benoit and G.W. Luther, *Environ. Sci. Technol.*, 1999, **33**, 3021–3026.
34. I. Ciglenecki, D. Krznaric and G.R. Helz, *Environ. Sci. Technol.*, 2005, **39**, 7492–7498.
35. D.S.d. Santos, R.A. Alvarez-Puebla, O.N.O. Jr and R.F. Aroca, *J. Mater. Chem.*, 2005, **15**, 3045–3049.
36. J.R. Lead, W. Davison, J. Hamilton-Taylor and J. Buffle, *Aquat. Geochem.*, 1997, **3**, 213–232.
37. J.R. Lead and K.J. Wilkinson, in *Environmental Colloids and Particles: Behaviour, Separation and Characterisation*, ed. K.J. Wilkinson and J.R. Lead, John Wiley and Sons, Chichester, 2006, vol. 10.
38. O. Gustafsson and P.M. Gschwend, *Limnol. Oceanogr.*, 1997, **42**, 519–528.
39. J. Buffle and G.G. Leppard, *Environ. Sci. Technol.*, 1995, **29**, 2169–2175.
40. F.R. Doucet, J.R. Lead and P.H. Santschi, in *Environmental Colloids: Behaviour, Structure and Characterisation*, ed. K.J. Wilkinson and J.R. Lead, John Wiley and Sons, Chichester, 2006, vol. 10.
41. J.R. Lead, W. Davison, J. Hamilton-Taylor and M. Harper, *Geochim. Cosmochim. Ac.*, 1999, **63**, 1661–1670.
42. B. Lyven, M. Hassellov, D.R. Turner, C. Haraldsson and K. Andersson, *Geochim. Cosmochim. Ac.*, 2003, **67**, 3791–3802.
43. P.G.C. Campbell, in *Metal Speciation and Bioavailability in Aquatic Systems*, ed. A. Tessier, Wiley, New York, 1995, pp. 45–102.
44. P.R. Paquin, J.W. Gorsuch, S. Apte, G.E. Batley, K.C. Bowles, P.G.C. Campbell, C.G. Delos, D.M.D. Toro, R.L. Dwyer, F. Galvez, R.W. Gensemer, G.G. Goss, C. Hogstrand, C.R. Janssen, J.C. McGeer, R.B. Naddy, R.C. Playle, R.C. Santore, U. Schneider, W.A. Stubblefield, C.M. Wood and K.B. Wu, *Comp. Biochem. Phys. C*, 2002, **133**, 3–35.
45. V.I. Slaveykova and K.J. Wilkinson, *Environ. Chem.*, 2005, **2**, 1–16.

Nanoparticles in the Atmosphere

ROY HARRISON

1 Introduction

Currently, the main exposures of the general population to nanoparticles are likely to arise from breathing the air. Nanoparticles are very abundant in the atmosphere and have been present since time immemorial in the vicinity of any combustion process. In most areas with heavy human occupancy, concentrations are highly elevated above unpolluted background concentrations and major exposures are unavoidable. There have been frequent suggestions that because of their enhanced toxicity per unit mass, nanoparticles are responsible for a large component of the adverse effects due to exposure to airborne particles. However, nanoparticles, although dominating the number concentration of particles in the atmosphere, represent only a relatively small proportion of the mass of PM_{10} or $PM_{2.5}$, and the available epidemiological evidence does not present a compelling case for indicting them as the major contributor to airborne particle toxicity.

2 Sources of Atmospheric Nanoparticles

2.1 Primary Emissions

In common with air pollutants in general, nanoparticles can be either primary or secondary. The term primary refers to those that are directly emitted, such as from road traffic exhaust and industrial combustion processes. The United Kingdom has developed a national inventory of emissions of nanoparticles (expressed as the mass of particles less than 0.1 μm diameter or $PM_{0.1}$) and the inventory for 1998[1] appears in Figure 1. It may be seen that this is dominated by emissions from road traffic (62%) with combustion processes, in industry, commercial, institutional or residential premises, energy production and transformation or waste treatment and disposal being the second most major

Issues in Environmental Science and Technology, No. 24
Nanotechnology: Consequences for Human Health and the Environment
© The Royal Society of Chemistry, 2007

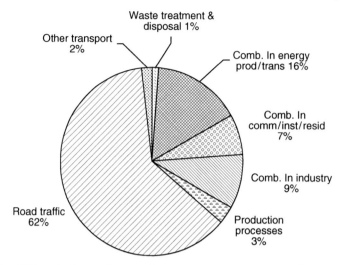

Figure 1   UK emissions of $PM_{0.1}$ by sector (ktonnes) (1998).[1]

contributor.[1] When compared to the national atmospheric emissions inventory for PM_{10}, the $PM_{0.1}$ inventory gives much greater emphasis to road traffic and other combustion sources. It is also notable that inventories of urban emissions of particulate matter show a much greater proportion coming from road traffic than is shown in the national emissions (due to the importance of road vehicles as a source of air pollution in urban areas), and if that factor is taken into account then it would be anticipated that road traffic emissions dominate the primary nanoparticles in the urban atmosphere and indeed measurement studies tend to confirm that impression.[2]

2.2   Secondary Particles

In addition to primary emitted particles there are also secondary particles in the atmosphere. Such particles are formed within the atmosphere itself from the condensation of low volatility vapours formed from the oxidation of atmospheric gases. Wholly new particles can be formed by a process known as homogeneous nucleation and relatively recent work has shown that in the presence of small concentrations of ammonia, sulfuric acid formed from the oxidation of sulfur dioxide is able to nucleate with water to form new particles which are approximately 1–2 nm in diameter when formed.[3] These sulfate nuclei grow relatively rapidly[4] (often at more than 10 nanometres in diameter per hour) and it is believed that the growth is mainly due to the condensation of the oxidation products of volatile organic compounds. Urban air typically contains a high background level of pre-existing particles whose large surface area acts as a condensation sink for low volatility gases and therefore formation of new particles by nucleation in the urban atmosphere is a relatively un favourable process. A detailed study in Birmingham, UK, showed that nucleation processes could be demonstrated on only 3.4% of a total of 232 days spread through the year.[5] The most important common factor in determining the occurrence of nucleation was the low background of pre-existing particles.

Making measurements in Atlanta, USA, Woo *et al.*[6] observed frequent (23 days during April and August) events where pronounced peaks in the 3–10 nm size range occurred, typically around noon. The events occurred at times of elevated sulfur dioxide and low NO_x, and were attributed to nucleation processes. A second type of event involved elevated concentrations in the 10–35 nm and 35–45 nm size ranges during early morning and late afternoon, accompanied by elevated SO_2 and NO_x. It was suggested that a fossil fuel combustion source might have been responsible for the second type of event.

The sulfate nucleation and organic component growth model outlined above is not the only atmospheric nucleation process observed in the atmosphere. Some of the highest concentrations of airborne nanoparticles have been observed on the western (Atlantic) coast of Ireland in very clean air.[7] The common factor in determining high nanoparticle concentrations appears to be conditions of low tide during daytime, in which large seaweed beds are exposed to the atmosphere, releasing organoiodine compounds, which oxidise to form condensable species. Under such conditions airborne concentrations of nanoparticles well in excess of 10^5 per cubic centimetre are regularly observed. O'Dowd *et al.*[8] propose a mechanism in which CH_2I_2 released from macroalgae undergoes photolysis to iodine atoms, which form a range of iodine oxide species which condense to form an aerosol. Laboratory studies by Burkholder *et al.*[9] confirm that this mechanism is a feasible explanation.

In the terrestrial environment, there have been frequent observations of large number concentrations of nanoparticles forming over the boreal forests of Finland. O'Dowd *et al.*[10] have shown that the composition of 3–10 nm particles is largely organic and consistent with *cis*-pinonic or pinic acid, derived from the oxidation of biogenic terpenes. Laboratory studies by Zhang *et al.*[11] confirm that organic acids can co-nucleate with sulfuric acid promoting efficient formation of organic and sulfate aerosols. Kulmala *et al.*[4] have reviewed published observations of atmospheric formation of nanoparticles. Formation rates of 3 nm particles are in the range $0.01–10$ cm^{-3} s^{-1} in the rural atmospheric boundary layer, up to 100 cm^{-3} s^{-1} in urban areas, and as high as $10^4–10^5$ cm^{-3} s^{-1} in coastal areas and industrial plumes. Typical growth rates are in the range 1–20 nm h^{-1} in mid-latitudes, and can be as low as 0.1 nm h^{-1} in polar regions.[4]

2.3 Formation of Nanoparticles During Diesel Exhaust Dilution

When systematic studies were initiated to determine the size distribution of particles in engine exhaust, it soon became clear that both the total number of particles and their size distribution depended on the conditions of dilution of the exhaust pipe output. By far the most important factor was the dilution ratio, which describes the volume of diluted exhaust divided by the volume of raw exhaust.[12] Studies in which road vehicles have been followed by instrumented laboratories indicate that in the atmosphere the mixing of vehicle exhaust rapidly creates dilution ratios of the order of 1000 or more. When these were used in laboratory dilution tunnel experiments, they caused a very marked downward shift in the mean size of emitted particles and an increase in the overall number of particles, as illustrated in Figure 2. Whilst the overall volume of particulate matter did not change significantly between low and high dilution ratios, at high dilution ratios a greater proportion of the particle volume was present as nanoparticles of 10–30 nm diameter as opposed to coarser particles. Subsequent research has shown that diesel engine exhaust contains two predominant components. The largest mass (but not number) of particles is comprised by particles with a core of elemental (graphitic) carbon formed within the combustion chamber of the engine itself. The mode in the number size distribution of such particles is typically in the 30–100 nm size range. The second most numerous part of the particle size distribution comprises particles generally within the range 10–30 nm, which are formed through the condensation of semi-volatile vapours during the dilution of the engine exhaust.[12,13] The available research indicates that the most likely source of material for such particles is engine oil vaporised in the combustion process.

Using mass spectrometric techniques, Schneider *et al.*[14] have made measurements of newly formed exhaust particles in both the laboratory and the field. They concluded from their data that particle formation depends upon sulfuric acid/water nucleation followed by condensation of involatile and semi-volatile organic compounds on pre-existing particles. Exhaust dilution conditions, fuel

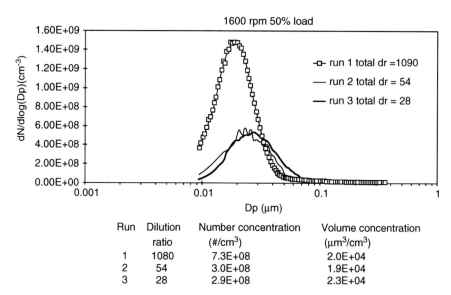

Run	Dilution ratio	Number concentration (#/cm^3)	Volume concentration (μm^3/cm^3)
1	1080	7.3E+08	2.0E+04
2	54	3.0E+08	1.9E+04
3	28	2.9E+08	2.3E+04

Figure 2 Comparison of particle size distribution measured at different dilution ratios at engine speed 1600 rpm and 50% load.[12]

sulfur content and engine load (which influences the conversion of fuel sulfur to sulfuric acid) were all found to be influential. In a second study, which sampled both in the laboratory and on the road, Rönkkö *et al.*[15] found that the nucleation mode, formed within 5 metres of the moving vehicle, did not change significantly with chase distance. Their results indicated that at low engine torques, hydrocarbons have an important role in the nucleation process, whilst at high torques the process appeared to be sulfur-driven.

3 Particle Size Distributions

If one considers a single particle of the size of a football and then imagines that particle broken down into many smaller particles, each the size of a pea, it can be readily appreciated that the same mass of material can comprise one very large particle or many thousands of much smaller particles. When the surface area of the smaller particles is summed, it is much greater than that of the original football-sized particle. This is expressed numerically in Table 1, in which the start point is a single particle of 10 µm diameter, which is then broken down into successfully smaller particles such that when it has been broken down into particles of 0.01 µm (or 10 nm) diameter, it has produced 10^9 particles with an increase of a factor of 10^6 in surface area. Figure 3 shows a particle size distribution measured in the atmosphere of Birmingham expressed in terms of number, surface area and particle volume. When the number of particles is expressed as a function of particle size, it is seen that a very large proportion of the particles fall into the nanoparticle size range of less than 0.1 µm diameter. However, when added together, such particles make up a very small proportion of the volume of particles shown in the lowest part of the graph in which the bimodal distribution typical of airborne particles is seen. This size distribution, which is fairly typical of a polluted urban atmosphere, demonstrates that nanoparticles make up by far the largest part of the number of airborne particles whilst contributing very little to volume (or mass) of particles, and only a limited amount to the surface area, which may be important in relation to some of the impacts on human health. One consequence of this relationship between the size distributions is that whilst being difficult to quantify by mass, ultrafine particles are easily measured by number. The total number count of particles in ambient air is typically dominated by those in the ultrafine fraction, and measurements of particle number are commonly used as a surrogate for ultrafine particle abundance.[16]

Table 1 Influence of particle size on particle number and surface area for a given particle mass.

Particle diameter	Relative number of particles	Relative surface area
10 µm	1	1
1 µm	10^3	10^2
0.1 µm	10^6	10^4
0.01 µm	10^9	10^6

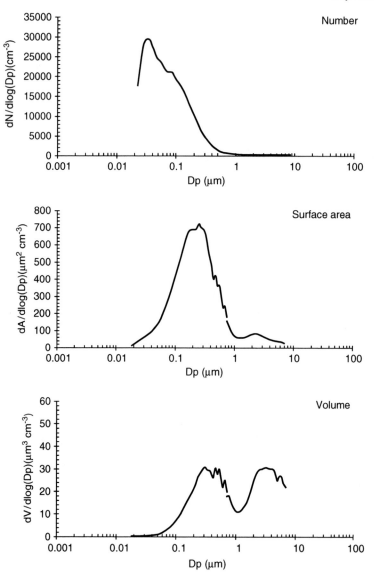

Figure 3 Particle size distribution measured in Birmingham.

3.1 Source Strength of Traffic Particles

Various workers have sought to quantify the particle emissions from vehicles driving on the road. Such measurements have included the fine ($PM_{2.5}$) fraction, assumed to arise mainly from engine exhaust, the coarse ($PM_{2.5-10}$) fraction assumed to represent mainly abrasion sources (tyre, brake and road surface wear) and resuspended road dust, and the number of particles. Two main methods have been used – calculation from either concentration measurements made in tunnels with known ventilation characteristics, or by ratio to another component such as NO_x for which the emission rate is known with some degree of confidence. The results of such studies, recently reviewed by Jones and Harrison[17] typically reveal $PM_{2.5}$ emission factors of the order of 0.2–1.0 g km^{-1} for heavy-duty and 0.01–0.03 g km^{-1} for light-duty vehicles, although these are inevitably speed and driving mode dependent. Emission factors for particle number range considerably from around $(2–7) \times 10^{13}$ km^{-1} for light-duty vehicles for ca. 10^{15} km^{-1} for heavy duty vehicles, dependent upon vehicle speed. Whilst there appears to be a general tendency for the emission factor to increase with vehicle speed, there are major differences between studies, which cannot be explained by this factor alone.

3.2 Emissions from Non-Traffic Sources

There is a paucity of data concerning emissions of ultrafine/nanoparticles from other combustion and high-temperature sources. Shi *et al.*[2] observed enhanced concentrations of nanoparticles in air impacted by emissions from local combustion sources, but atmospheric observations are few, and difficult against the generally high pre-existing background. Chang *et al.*[18] used a pilot combustion plant to study emissions of particles from burning coal, oil and gas in a stationary facility. As with diesel engines, the measured size distributions were sensitive to dilution conditions between the combustor and the analyser. The size distributions stabilised at dilution ratios above 50. The modal size and total number concentrations per unit combustion exhaust appear in Table 2. Whilst the size distributions showed a clear dependence upon aging time prior to sizing, the influence of fuel type was larger. As Table 2 illustrates, the modal particle size increases in the order natural gas < coal < fuel oil. Whilst the total number of particles is not very sensitive to the type of fuel, the particle mass will differ greatly with fuel oil > coal > natural gas within the range of measurements (10–420 nm).

4 Measurement of Nanoparticles in Roadside Air

Measurement techniques for nanoparticles are described in Chapter 4 in the context of the occupational environment, but are mostly applicable also to

Table 2 Approximate modal diameter (after dilution) and total number concentration in combustion exhaust for particles emitted from the combustion of coal, No. 6 fuel oil and natural gas[16].

	Coal	*Fuel Oil*	*Natural Gas*
Modal diameter (nm)	35	95	15
Number concentration (cm^{-3})	3.8×10^6	$1.1–1.2 \times 10^6$	$3.9–4.2 \times 10^6$

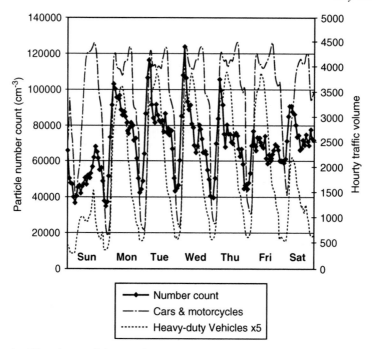

Figure 4 Hourly particle number count (CPC) and traffic volume averaged over the week.

outdoor air. Marylebone Road is a major highway in central London, which runs through a street canyon with traffic volumes exceeding 80 000 vehicles per day on a six-lane highway. Figure 4 shows an averaged weekly diurnal variation of particle number count measured with a condensation particle counter together with the volume of light-duty and heavy-duty vehicles. It may readily be seen that the closest relationship is between the particle number count and the flow of heavy-duty (diesel) vehicles. The curves do, however, differ particularly in that the particle number count peaks early in the morning and then tends to decline more rapidly than the traffic count. A close examination of the data showed that this was due to the smallest fraction of measured particles, those between 11 and 25 nm increasing in relative abundance with falling air temperature.[19] This is attributable to the colder air favouring rapid condensation and particle nucleation. Similarly, increases in windspeed increased the abundance of this particle size fraction relative to particles greater than 100 nm diameter, presumably because of the greater dilution and lower condensation sink.

The number distribution of particle sizes measured at Marylebone Road appears in Figure 5, the smallest size mode occurring in the nighttime samples. An hierarchical cluster analysis showed that the number of particles as well as the mass of particles was closely related to the NO_x and heavy-duty vehicle flow. On the other hand, the carbon monoxide was linked to the predominantly gasoline light-duty vehicle fleet.[20]

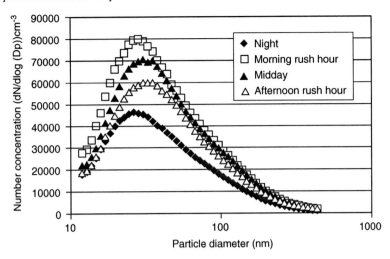

Figure 5 Particle size distributions measured at Marylebone Road, London.

5 Transformation and Transport of Ultrafine Particles

There is a number of processes which nanoparticles can undergo once present in the atmosphere. As noted above, they can grow by condensation of low volatility components, but equally if the vapour phase concentrations are low they can shrink by evaporation. This leads to a change in the size distribution but not the overall number concentration. On the other hand, coagulation, in which two particles collide and stick, which is slow except at high number concentrations, leads to a reduction in the particle number concentration but an increase in the particle size. Particles also are lost from the atmosphere by dry and wet deposition, both of which are quite efficient for very small particles. This leads to a reduction in particle number concentration and a shift in the size distribution. One of the most important processes for emitted nanoparticles is dilution with cleaner air. Thus, for example, particles from traffic mix upwards with less polluted air leading to a reduction in the number concentrations and generally a shifting of the size distribution towards larger sizes simply because the dilution air contains a distribution of larger particle sizes. Various researchers working close to major highways or with models of particle behaviour in the urban atmosphere have drawn different conclusions concerning the predominant processes. Zhang et al.[21] interpreted the evolution of size distributions near roadways as indicating that condensation/evaporation and dilution are the dominant processes. Gidhagen et al.,[22] looking again at the evolution of particle sizes near roadways, saw little effect on coagulation but a loss of about 10% of particles by dry deposition. Clarke et al.[23] ran a Lagrangian model of aerosol transport over London concluding that the dominant process is upward dispersion. Coagulation and deposition were found to have little effect upon the size distribution.

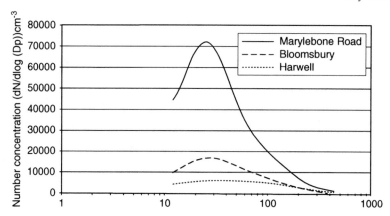

Figure 6   Particle number size distribution at three sites (April & May 1998).

One consequence of these atmospheric processes, which is seen in Figure 6, is that average particle size distributions measured over the same time period show a substantial change between the urban street canyon at Marylebone Road, which is dominated by fresh vehicle emissions, the urban background at Bloomsbury, at which much of the aerosol has undergone some processing and the rural background at Harwell, which shows a distribution much more typical of the aged accumulation mode of particles.

The semi-volatility of the nanoparticle mode is underlined not only by the formation process but also by the fact that studies of particle volatility demonstrate particle shrinkage. Thus, for example, Kuhn *et al.*[24] examining the volatility of particles collected near a US freeway showed that particles of 27 nm shrank to a mode of 24 nm at 60°C and to 15 nm at 110°C. Particles of 90 nm shrank to a mode of 80 nm at 60°C and 60 nm at 110°C. In terms of volume, these particles had shrunk to 17% and 30% respectively of their original volumes when heated to 110°C.

6   Measurements of Particle Number Concentration in the Atmosphere

The UK has a network of continuous particle counting instruments, which are able to determine the number concentrations integrated across all particle sizes greater than 7 nm diameter. This network has demonstrated strong gradients between the relatively low concentration at Port Talbot, which has influence from a steelworks and (at some distance) a motorway, urban centres of Glasgow, London, Birmingham, Manchester and Belfast and the street canyon at Marylebone Road, which shows by far the highest number concentrations.[25] A seasonal pattern with higher concentrations in the winter and lower in the summer is probably reflective of both less efficient nucleation and more efficient atmospheric mixing during the hotter months of the year (Figure 7). The

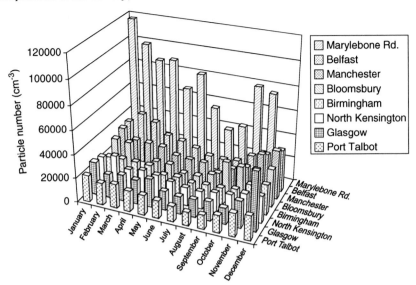

Figure 7 Seasonal influence on number concentration at various UK sites. [23]

diurnal profile of particle concentrations at urban centres very much reflects that typical of traffic-generated pollutants in which there is a morning rush-hour peak followed by a gentle decline in concentrations which then rise more gently for a secondary peak corresponding to the evening rush hour. Concentrations then decline slowly throughout the night to a minimum at around 4 am. Jones and Harrison[25] examined the relationships between particle number count measured in central London and the daily PM_{10} concentration at the same location. They found a distinct seasonal pattern in the results whereby in the summer months particle number concentrations rarely exceeded 20 000 cm^{-3} despite PM_{10} concentrations reaching in excess of 100 $\mu g\ m^{-3}$. On the other hand, in the winter months, high PM_{10} concentrations were typically associated also with high particle number concentrations of up to 80 000 cm^{-3}. They interpreted this pattern of behaviour as reflecting the fact that long range transported secondary aerosols are generally responsible for high PM_{10} concentrations in the summer months, whereas in the winter months there are frequent events caused by trapping of local traffic-generated pollutants, in which case both number concentrations and PM_{10} mass would be high.

7 Chemical Composition of Atmospheric Nanoparticles

Cass and co-workers[26] have published on measurements of the chemical composition of particles in the size range smaller than 100 nanometres sampled in California. They were able to account for a very large percentage of the measured mass with a typical set of results appearing in Table 3. This shows that organic compounds comprise by far the major component of ultrafine

Table 3 Two examples of composition measurements of atmospheric nano-
            particles[24].

	Riverside, 1996	*Azusa, California, 1997*
Organic compounds	67.2%	51.9%
Elemental Carbon	3.8%	4.9%
Ammonium	1.3%	9.3%
Nitrate	2.8%	1.3%
Sulfate	8.6%	12.0%
Chloride	0.9%	0.1%
Sodium	0.3%	0.4%
Metal oxides	12.0%	20.2%

particles consistent with their origin in traffic (mainly diesel) exhaust emissions. Lesser but very significant contributors are metal oxides, which probably arise primarily from fuel impurities and engine wear together with lesser amounts of elemental carbon (formed in the engine), ammonium and sulfate and very trace amounts of nitrate, chloride and sodium. A more recent study by Sardar *et al.*[27] has shown broadly similar findings. Sampling in urban source regions and inland receptor sites in California, Sardar *et al.*[27] measured ultrafine mass concentrations in autumn. Predominant chemical components were organic carbon (32–60%) and nitrate (0–4%). Clearly, in non-urban locations, the composition of nanoparticles may differ considerably, according to the nature of local sources.

Lin *et al.*[28] found Aitken size modes for a number of trace metals sampled beside a heavily trafficked road, which they attributed to the local origins of the particles. The Ag, Cd, Cr, Ni, Pb, Sb, V and Zn were 37, 50, 28, 30, 24, 64, 38 and 22% by mass respectively present in ultrafine (< 0.1 μm) particles. Harrison *et al.*[29] found important abundances of Be, Pb, Cd, Cs, Bi, Se, Ni, Mo, Ca and Zn in particles of < 0.2 μm in the roadside environment. Cu, Ba and Ca showed correlation with NO_x, and Ca with 4, 5-dimethylphenanthrene confirming the automotive source of these nanosized particulate metals.

8 Indoor/Outdoor Relationships of Nanoparticles

A number of studies have measured the indoor concentrations of nanoparticles and some have studied indoor/outdoor relationships. Generally, the number count of particles indoors is for much of the time smaller than that outdoors, reflecting the infiltration of outdoor particles with losses both during infiltration and also by deposition once inside the building. However, all studies have shown intermittent peaks of ultrafine particles in occupied buildings associated with indoor activities. Such peaks can vastly exceed the measurements taken out-of-doors (see Figure 8). The large excursions are most typically connected with cooking activities, which comprise a very major source of ultrafine particles within the home. When these peaks in concentration associated with indoor sources are edited out of the dataset, indoor and outdoor concentrations

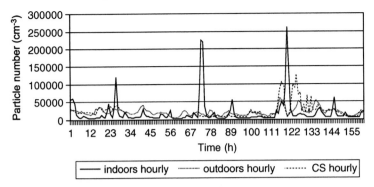

Figure 8 Time series of indoor, outdoor and central site particle number concentrations.[26]

are generally quite well correlated with indoor levels at about 20% of those out-of-doors.[30]

Matson[31] compared indoor and outdoor concentrations of nanoparticles in rural and urban areas of Scandinavia showing typical indoor/outdoor ratios of between 0.5 and 0.8. Ratios exceeding unity were observed in offices due to smoking, and in residential buildings due to activities such as cooking and candle burning. Working in Finland, Hussein *et al.*[32] found a mean value of indoor/outdoor ratio of 0.36 for particles of < 100 nm diameter. A lag was seen between outdoor and indoor concentrations. The loss rate coefficient of ultra-fine particles in indoor air ranged from 2 h^{-1} for 10 nm particles to 0.1 h^{-1} for 100 nm diameter particles. Zhu *et al.*,[33] working near a California freeway with air exchange rates varying from 0.3–1.11 h^{-1}, found indoor/outdoor ratios to be largest (0.6–0.9) for particles of 70–100 nm, whilst the lowest values (0.1–0.4) were for particles of 10–20 nm.

Indoor sources can make a major contribution to indoor concentrations of nanoparticles. Dennekamp *et al.*[34] found both gas and electric cooking to be a substantial source of nanoparticles. The higher concentrations of particles were generally from gas cooking, with modal diameters ranging from 16–69 nm. Wallace,[35] working in an occupied townhouse, found a gas clothes drier to be a major particle source, contributing a source strength estimated as above 6×10^{12} particles per drying episode, resulting in peak number concentration in the room exceeding 10^5 cm^{-3}.

9 Conclusions

Knowledge of the sources and concentrations of nanoparticles is relatively good in two areas, *i.e.* vehicle emissions and atmospheric formation events through nucleation. This knowledge is primarily of number concentrations, and far less is known of particle chemistry. There are few data regarding nanoparticles from other combustion processes, and virtually none on engineered nanoparticles in the atmosphere. Currently, there are no quick, effective means of distinguishing engineered nanoparticles from the substantial background concentrations deriving from other sources.

References

1. AQEG, 2005, Department of Environment, Food and Rural Affairs, London.
2. J.P. Shi, D.E. Evans, A.A. Khan and R.M. Harrison, *Atmos. Environ.*, 2001, **35**, 1193.
3. M. Kulmala, *Science*, 2003, **302**, 1000.
4. M. Kulmala, H. Vehkamäki, T. Petäjä, M. Dal Maso, A. Lauri, V.-M. Kerminen, W. Birmili and P.H. McMurry, *Aerosol Science*, 2004, **35**, 143.
5. A. Alam, J.P. Shi and R.M. Harrison, *J. Geophys. Res.*, 2003, **108**, 4093 (2003).
6. K.S. Woo, D.R. Chen, D.Y.H. Pui and P.H. McMurry, *Aerosol Sci. & Technol.*, 2001, **34**, 75.
7. A.G. Allen, J.L. Grenfell, R.M. Harrison, J. James and M.J. Evans, *Atmos. Res.*, 1999, **51**, 1.
8. C. O'Dowd, J.L. Jimenez, R. Bahrenini, R.C. Flagan, J.H. Seinfeld, K. Hämeri, L. Pirjola, M. Kulmala, S.G. Jennings and T. Hoffmann, *Nature*, 2002, **417**, 632.
9. J.B. Burkholder, J. Curtius, A.R. Ravishankara and E.R. Lovejoy, *Atmos. Chem. Phys.*, 2004, **4**, 19.
10. C.D. O'Dowd, P. Aalto, K. Hämeri, M. Kulmala and T. Hoffmann, *Nature*, 2002, **416**, 497.
11. R. Zhang, I. Suh, J. Zhao, D. Zhang, E.C. Fortner, X. Tie, L.T. Molina and M.J. Molina, *Science*, 2004, **304**, 1487.
12. J.P. Shi and R.M. Harrison, *Environ. Sci. Technol.*, 1999, **33**, 3730.
13. J.P. Shi, D. Mark and R.M. Harrison, *Environ. Sci. Technol.*, 2000, **34**, 748.
14. J. Schneider, N. Hock, S. Weimer and S. Borrmann, *Environ. Sci. Technol.*, 2005, **39**, 6153.
15. T. Rönkkö, A. Virtanen, K. Vaaraslahti, J. Keskinen, L. Pirjola and M. Lappi, *Atmos. Environ.*, 2006, **40**, 2893.
16. R.M. Harrison, J.P. Shi, S. Xi, A. Khan, D. Mark, R. Kinnersley and J. Yin, *Phil. Trans. R. Soc. Lond. A*, 2000, **358**, 2567.
17. A.M. Jones and R.M. Harrison, *Atmos. Environ.*, submitted.
18. M.-C.O. Chang, J.C. Chow, J.G. Watson, P.K. Hopke, S.-M. Yi and G.C. England, *J. Air & Waste Manage. Assoc.*, 2004, **54**, 1494.
19. A. Charron and R.M. Harrison, *Atmos. Environ.*, 2003, **37**, 4109.
20. A. Charron and R.M. Harrison, *Environ. Sci. & Technol.*, 2005, **39**, 7768.
21. K.M. Zhang, A.S. Wexler, D.A. Niemeier, Y.F. Zhu, W.C. Hinds and C. Sioutas, *Atmos. Environ.*, 2005, **39**, 4155.
22. L. Gidhagen, C. Johansson, G. Omstedt, J. Langner and G. Olivares, *Environ. Sci. Technol.*, 2004, **38**, 6730.

23. A.G. Clarke, L.A. Robertson, R.S. Hamilton and B. Gorbunov, *Sci. Total Environ.*, 2004, **334–335**, 197.
24. T. Kuhn, M. Krudysz, Y. Zhu, P.M. Fine, W.C. Hinds, J. Froines and C. Sioutas, *J. Aerosol Science*, 2005, **36**, 291.
25. R.M. Harrison and A.M. Jones, *Environ. Sci. Technol.*, 2005, **39**, 6063.
26. G.R. Cass, L.A. Hughes, P. Bhave, M.J. Kleeman, J.O. Allen and L.G. Salmon, *Phil. Trans. R. Soc. Lond. A*, 2000, **358**, 2581.
27. S.B. Sardar, P.M. Fine, P.R. Mayo and C. Sioutas, *Environ. Sci. Technol.*, 2005, **39**, 932.
28. C.-C. Lin, S.-J. Chen and K.-L. Huang, *Environ. Sci. Technol.*, 2005 **39**, 8113.
29. R.M. Harrison, R. Tilling and M.S. Callén Romero, *Atmos. Environ.*, 2003, **37**, 2391.
30. R.M. Harrison, C. Meddings, S. Thomas and J.M. Delgado Saborit, in *Proceedings of the Eighth Annual UK Review Meeting on Outdoor and Indoor Air Pollution Research*, 2004, **No. 5.12**, http://www.le.ac.uk/ieh/publications/publications.html.
31. U. Matson, *Sci. Total Environ.*, 2005, **343**, 169.
32. T. Hussein, K. Hämeri, M.S.A. Heikkinen and M. Kulmala, *Atmos. Environ.*, 2005, **39**, 3697.
33. Y. Zhu, W.C. Hinds, M. Krudysz, T. Kuhn, J. Froines and C. Sioutas, *J. Aerosol Science*, 2005, **36**, 303.
34. M. Dennekamp, S. Howarth, C.A.J. Dick, J.W. Cherrie, K. Donaldson and A. Seaton, *Occup. Environ. Med.*, 2001, **58**, 511.
35. L. Wallace, *Atmos. Environ.*, 2005, **39**, 5777.

Occupational Exposure to Nanoparticles and Nanotubes

DAVID MARK

1 Introduction

A discussion of the exciting progress that is rapidly taking place in the development and application of new nanomaterials is presented elsewhere in this publication (Chapter 1). In addition, the toxicological properties of nanoparticles and the potential human health effects of nanoparticle exposure are discussed in Chapters 5 and 6. The aims of this chapter are to outline the likely routes of exposure to humans in the workplace, discuss the methods available for assessing the level of exposure and to present a summary of currently published exposure data for nanoparticles.

There have been a number of recent reviews of the potential risks to human health of the manufacture and use of nanoparticles.[1-6] In the seminal report published by the UK Royal Society and Royal Academy of Engineering,[1] it is suggested that the population currently most at risk from nanoparticles is that of researchers and scientists in universities and research laboratories who develop new nanomaterials and/or devise uses for them for a wide range of applications. As these developments are scaled up and go into full-scale production, the emphasis will shift to those workers operating those processes, and handling and using the nanoparticle powders and products. Subsequently, in the life cycle of the material, consumers may be exposed if the nanoparticles that are incorporated into products such as surface coatings, plastics, etc.,[7] are worked (sawn, machined, sanded, etc.) or degraded (mechanically or by water, sun, chemicals, etc.) so that they become free and available for exposure to humans. Nanoscale particles are currently included in a number of consumer products such as sunscreens and cosmetics, and for these the potential for dermal penetration of the particles leading to ill health has raised concern.

This chapter concerns the assessment of exposure of workers as a result of the manufacture, handling and use of nanoparticles. The scientific framework

Issues in Environmental Science and Technology, No. 24
Nanotechnology: Consequences for Human Health and the Environment
© The Royal Society of Chemistry, 2007

for exposure assessment is discussed with a consideration of the routes of exposure and the most appropriate measurement metric to be used. This is followed by a summary of the instrumentation currently available and a discussion of possible sampling strategies. A review of currently published data on workplace measurements is then presented and the chapter ends with a discussion of current research projects that will hopefully provide a large enough database to enable risk assessments to be carried out with confidence.

For simplicity in this chapter, the term "nanoparticles" will include nano-tubes, unless their particular morphology (fibre-like shape) requires that they should be considered separately (*e.g.* in exposure assessment).

2 Scientific Framework for Assessing Exposure to Nanoparticles

2.1 Terminology and Definitions

Nanotechnology encompasses a wide range of technologies and disciplines and the production and use of nanoparticles is but a small area. It is defined in the BS Publicly Available Specification PAS 71[8] as "the design, characterisation, production and application of structures, devices and systems controlling shape and size with one or more dimensions of the order of 100 nm or less". This wide-ranging document includes definitions for over 150 terms relating to areas such as particle names and characteristics, production methods and measure-ment methods. For terms of specific relevance to workplace health and safety issues of nanoparticles a more informative document is the ISO Technical Report "Workplace atmospheres – Ultrafine, nanoparticle and nano-structured aerosols – Inhalation exposure characterization and assessment".[9] In this document a *nanoparticle* is defined as "a particle with a nominal diameter (such as geometric, aerodynamic, mobility, projected-area or otherwise) smaller than about 100 nm". An *ultrafine particle* is defined similarly and is commonly used interchangeably with nanoparticle, but is often used in the context of particles produced as a by-product of a process (incidental particles), such as welding fume and combustion fume and to describe ambient particles outdoors that are smaller than 100 nm. Larger particles with a nanometre-scale structure (such as agglomerates of nanoparticles and nanometre-diameter fibres) are referred to as *nanostructured* particles, and aerosols of nanoparticles and nanostructured particles are referred to as *nanoaerosols*.

The terms *engineered nanoparticle* and *engineered nanoaerosol* have also been used loosely to describe particles and aerosols associated with engineered nanometre-structured materials that have been intentionally engineered and produced with specific properties.

2.2 Routes of Exposure

There are three main routes by which workers can be exposed to nanoparticles: a) inhalation, b) ingestion and c) dermal penetration.[10]

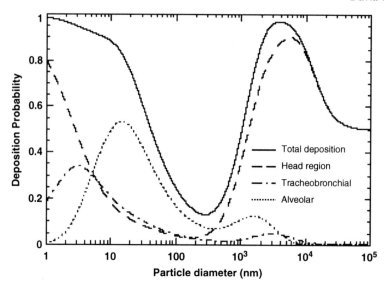

Figure 1   Predicted total and regional deposition of particles in the human
respiratory tract related to particle size using ICRP 66 model.
Deposition fraction includes the probability of particles being inhaled (inhalability). The subject is considered to be a nose breather,
performing standard work.

Inhalation: As with most particles in the workplace, inhalation is considered
to be the main route by which free unbound nanoparticles will enter the bodies
of workers. Once inhaled, nanoparticles will deposit in all regions of the
respiratory tract with high efficiency, dependent upon their particle size. Figure 1
shows the fractional deposition of inhaled nanoparticles in the nasopharyngeal,
tracheobronchial and alveolar regions of the human respiratory tract for nasal
breathing at rest using the predictive mathematical model of the ICRP.[11]

It is interesting to note that 90% of the 1 nm particles are deposited in the
nasopharyngeal region, with 10% in the tracheobronchial region and none in
the alveolar region. For 20 nm particles, however, 50% deposit in the alveolar
region and 25% in the nasopharyngeal and tracheobronchial regions. Once
deposited, nanoparticles, in contrast to larger particles, appear to translocate to
different organs in the body after penetrating the cell epithelium and entering
the blood or lymph systems. Consequently, this will be the main route of
exposure to be discussed in this chapter.

Ingestion: In the workplace, the main routes by which nanoparticles can be
ingested are by swallowing the mucous that traps the particles deposited in the
airways and moves them up to the nasopharyngeal area where it is swallowed
or expectorated, or by sucking or licking a contaminated surface. Nanoparticles are also used in foods and drugs that are swallowed, but for these exposure
cannot be considered to be as a result of work practices. Only a few studies have

been carried out to investigate the uptake and disposition of nanoparticles to the GI tract and most have shown that they pass through the GI tract and are eliminated quickly.[10]

Dermal penetration: Nanoparticles of titanium dioxide and zinc oxide are currently used in advanced sunscreens to enhance the UV absorption efficiency and because they become transparent. The evidence of potential penetration into the epidermis, however, is mixed and so they continue to be used. In the workplace, the main potential cause by which the skin can be exposed to nanoparticles is by handling nanopowders during their manufacture or use. It is still unclear whether nanoparticles will penetrate the skin and cause any toxicological problems. However, most of the limited work that has been reported has been carried out on intact skin. The effect of flexing the skin has yet to be fully explored as has the penetration through damaged skin.[12]

2.3 Metric to be used for Assessing Exposure to Airborne Nanoparticles

The current method of assessing worker exposure to airborne particles in workplaces involves the measurement of the mass concentration of health-related fractions of particles in the worker's breathing zone[13] and their chemical composition. The health-related aerosol fractions defined relate to the probability of penetration of airborne particles to the various anatomical regions of the respiratory system and provide a specification for the performance of

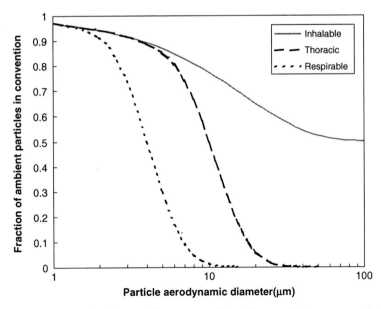

Figure 2 Health-related sampling conventions for workplace aerosols from IS 7708[13].

sampling instruments (see Figure 2). The inhalable convention is the mass fraction of total airborne particles that enters the nose or mouth during breathing, The thoracic convention is the mass fraction of inhaled particles that penetrate the larynx, with 50% penetration at 11.64 μm (equivalent to 10 μm when expressed as a fraction of total aerosol) and the respirable convention is the fraction of inhaled particles that penetrate to the alveolar region of the lung, with 50% penetration at 4.25 μm (equivalent to 4 μm when expressed as a fraction of total aerosol).

The main exceptions to this methodology are a) for fibres such as asbestos, for which a downward facing cowled sampler collects particles on a filter and the respirable particles (defined as of diameter < 3 μm and aspect ratio > 3:1) are counted using optical or electron microscopy,[14] and b) for microorganisms, for which the standard method is to collect them on a growth medium and to count the number of colony-forming units.[15]

However, recent toxicological evidence is indicating that the potential health effects associated with inhaling nanoaerosols may not be closely associated with particle mass. A number of studies have indicated that the toxicity of insoluble materials increases with decreasing particle size, on a mass-for-mass basis.[10,16,17] The mechanisms by which these materials exhibit higher levels of toxicity at smaller particle sizes have yet to be explained, although there are many hypotheses. A number of studies indicate that biological response depends on the surface area of particles deposited in the lungs.[10,18,19] It has also been suggested that due to their small diameter, nanoparticles are capable of penetrating epithelial cells, entering the bloodstream from the lungs,[20] and even entering the brain via the olfactory nerves.[21] Now as particles in the nanometre size range have a high percentage of surface atoms, and are known to show unique physico-chemical properties, it would be expected that nanoparticles would demonstrate biological behaviour closely associated with particle diameter, surface area and surface activity.

It is apparent from the above discussion that measuring exposures to nanoaerosols in terms of mass concentration alone is not sufficient to assess potential health risk. In addition, there is strong evidence to suggest that occupational nanoaerosols should be monitored with respect to surface area. However, in this context, aerosol surface area is not well defined and it is dependent on the measurement method used. Geometric surface area refers to the physical surface of an object, and is dependent on the length-scale used in the measurement. Measurement length-scale determines the upper size of features that are not detected by the measurement method. For example, methods utilising molecular surface-adsorption have a length-scale that approximates to the diameter of the adsorbed molecules.[22] Similarly, biologically relevant surface area will most likely be determined by the smallest biological molecule that interacts with particles within the body.

Whilst a strong case may be made for using aerosol surface area as an exposure metric, it is also necessary to consider characterising exposures against aerosol mass and number concentration until further information is available. For each of these exposure metrics, but particularly in the case of mass concentration, particle size selective inlets will need to be employed to ensure only particles within the relevant size range are sampled.

The actual cut size that particle selection should be made for assessing potential human health impact is still open to debate and depends upon particle behaviour and subsequent biological interactions. The currently proposed cut size for nanoparticles is 100 nm, although this is not derived from particle behaviour in the respiratory tract following deposition. It is possible to develop a health-related definition of a nanoparticle based on deposition probability in the lungs (see the curves in Figure 2). Rather it comes from the changes in physical properties that some materials undergo below particle diameters of 200 nm to 300 nm. As particles become smaller, surface curvature, the arrangement (and percentage) of atoms on the particle surface and size-dependent quantum effects such as quantum confinement play an increasingly significant role in determining behaviour.

The main problem with using a single cut size of 100 nm is whether to consider nanostructured particles that are larger than 100 nm but have a sub-100 nm structure, such as agglomerates of nanoparticles with low fractal dimensions. It is currently unclear whether the biological impact of discrete nanoparticles depositing within the respiratory system is distinct from or similar to the impact of large agglomerates or aggregates of nanoparticles containing the same volume of material. If agglomerates or aggregates of nanoparticles either de-agglomerate or disaggregate completely following deposition, it is conceivable that the resulting biological impact will be similar to an equivalent exposure of discrete nanoparticles. In addition, if biological response is associated with the surface area of the deposited aerosol, then for a given volume of material, the response to deposited agglomerates/aggregates with an open fractal-like structure will conceivably be similar to that from an equivalent dose of discrete particles. However, if the nanostructured particles do not de-agglomerate then it is likely that they will not translocate to other organs in the body as readily as the discrete nanoparticles and so the biological impacts will be different. So, knowledge of the ease with which the specific particles will de-agglomerate will be required before deciding at which particle size to exclude unwanted (non-nanostructured) particles and this may vary with the particle properties.

3   Review of Methods for Assessing Exposure to Nanoparticles

3.1   General

There are generally four main reasons for measuring levels of pollutants in workplaces:

(i) assessment of personal exposure for compliance with regulations

(ii) assessment of personal exposure for linking with potential adverse health effects in epidemiological studies

(iii) identification of major emission sources for establishing a targeted control plan

(iv) assessment of efficiency of control systems deployed.

Each of these tasks requires specific and often different types of instrumentation. For example, for personal exposure measurements it is generally necessary to use small, battery-powered samplers, mounted on the worker's body, that move with him/her during the working shift. For source identification, portable monitors can be used, generally giving continuous measurements of concentration that can be correlated with details of the location, ventilation and the specific work processes being undertaken. In order to assess the efficiency of control measures in the workplace, many different types of instruments can be used including static, mains-driven instruments, depending upon the information required.

However, for assessing exposure to engineered nanoparticles we have a major confounding factor in most workplaces – that of the soup of ultrafine particles found in ambient aerosols[23] that will penetrate into the workplace to differing degrees, dependent upon the ventilation arrangements at each workplace. Possible methods of discriminating between exposure to engineered nanoparticles and to ambient ultrafine particles will be discussed later.

Although there is increasing toxicity evidence to suggest that particle surface area and size or even particle number are more relevant parameters for assessing risk to health from inhaling nanostructured particles than particle mass, no decision has yet been made about which to choose as the metric for exposure assessment. Consequently, sampling campaigns that are currently being planned in studies such as the EU-funded research project NANOSAFE2[24] will use a wide range of equipment to take measurements of mass, number and surface area concentrations. In addition, nanoparticles will be collected for physical and chemical characterisation off-line, back in the laboratory.

A brief description of the equipment currently available for measuring concentrations in workplaces will now be given. Generally, the equipment for the occupational hygiene measurements mentioned above should be personal, or at least portable, battery-powered, robust and relatively inexpensive. However, until a health-related sampling convention is agreed for nanoparticles, no purpose-built device will be available. It is prudent, therefore, to include research instrumentation capable of providing size distribution measurements as mentioned above. A summary of the equipment together with comments is given in Table 1, which is an updated version of that found in the ISO Technical Report on Ultrafine, nanoparticle and nano-structured aerosols – Inhalation exposure characterization and assessment.[9] Inclusion of instruments in the table or in discussion does not imply any endorsement.

3.2   Mass Concentration

Mass concentration measurements provide a link with current sampling conventions and methodologies for health-related aerosols. It is expected that the

Table 1   Instruments and techniques for monitoring nanoaerosol exposure.

Metric	Devices	Remarks
Mass directly	Size selective static sampler	The only devices offering a cut-point around 100 nm are cascade impactors (Berner-type low-pressure impactors or Microorifice impactors). Allows gravimetric and chemical analysis of samples on stages below 100 nm.
	TEOM®	Sensitive real-time monitors such as the Tapered Element Oscillating Microbalance (TEOM) may be useable to measure nanoaerosol mass concentration on-line, with a suitable size selective inlet.
Mass by calculation	ELPI	Real time size-selective (aerodynamic diameter) detection of active surface-area concentration giving aerosol size distribution. Mass concentration of aerosols can be calculated, only if particle charge and density are assumed or known. Size-selected samples may be further analysed off-line (as above).
	SMPS	Real time size-selective (mobility diameter) detection of number concentration, giving aerosol size distribution. Mass concentration of aerosols can be calculated, only if particle shape and density are known or assumed.
Number directly	CPC	CPCs provide real time number concentration measurements between their particle diameter detection limits. Without a nanoparticle pre-separator, they are not specific to the nanometre size range. P-Trak has diffusion screen to limit top size to 1 μm.
	SMPS	Real time size-selective (mobility diameter) detection of number concentration, giving number-based size distribution.
	Electron microscopy	Off-line analysis of electron microscope samples can provide information on size-specific aerosol number concentration.
Number by calculation	ELPI	Real time size-selective (aerodynamic diameter) detection of active surface-area concentration, giving aerosol size distribution. Data may be interpreted in terms of number concentration. Size-selected samples may be further analysed off-line.

Table 1 (*continued*)

Metric	Devices	Remarks
Surface area directly	Diffusion charger	Real-time measurement of aerosol active surface area. Active surface area does not scale directly with geometric surface area above 100 nm. Note that not all commercially available diffusion chargers have a response that scales with particle active surface area below 100 nm. Diffusion chargers are only specific to nanoparticles if used with an appropriate inlet pre-separator.
	ELPI	Real time size-selective (aerodynamic diameter) detection of active surface area concentration. Active surface-area does not scale directly with geometric surface area above 100 nm.
	Electron microscopy	Off-line analysis of electron microscope samples can provide information on particle surface-area with respect to size. TEM analysis provides direct information on the projected area of collected particles, which may be related to geometric area for some particle shapes.
Surface area by calculation	SMPS	Real time size-selective (mobility diameter) detection of number concentration. Data may be interpreted in terms of aerosol surface area under certain circumstances. For instance, the mobility diameter of open agglomerates has been shown to correlate well with projected surface area.[3]
	SMPS and ELPI used in parallel	Differences in measured aerodynamic and mobility can be used to infer particle fractal dimension, which can be further used to estimate surface area.

mass concentrations of nanoparticles in workplaces will be low, at least in comparison with the corresponding respirable fraction. High sampling flow rates will therefore be required to collect sufficient material for subsequent analysis, and it is doubtful if even the best personal sampling pumps presently available (with a maximum flowrate of $10 \, l \, min^{-1}$–$15 \, l \, min^{-1}$) can be used other than in specific sampling situations. High flowrate mains-powered pumps will therefore be required and these are heavy, bulky and not easily portable. However, the sampling head itself may be relatively compact if it consists of a single stage pre-separator with a cut-size of 100 nm and a collection substrate. This will enable the sampling head to be positioned close to a worker's breathing zone in situations where the worker is stationary. At the present time, there are no commercially available workplace aerosol samplers with a 100 nm cut point, although it should be possible to design and test a suitable device. There are very few data on expected mass concentrations of nanoparticles in workplaces but it is expected that flowrates of about 100 l min^{-1} will be required to collect sufficient mass of nanoparticles in an 8-hour shift to be above the limit of quantification for weighing. At these high flowrates, it should be possible in principle to operate existing devices such as impactors and cyclones to provide a cut point at 100 nm.

An alternative approach that has been used in both workplace[25,26] and environmental[27] studies is to use a low-pressure cascade impactor (*e.g.* Berner-type LPI) or micro-orifice cascade impactor (*e.g.* MOUDI). Both of these devices use inertial impaction to separate particles into discrete fractions according to their aerodynamic diameters and have two or three stages in the nanoparticle size range. In both devices, the masses of nanoparticles can be assessed by weighing the collection substrates before and after sampling, plotting the full size distribution and making a cut at 100 nm or whatever particle size is considered to be relevant for nanostructured particles. These will be described in more detail later.

In a recent study of mass, number and surface area concentration relationships carried out by Wake,[28] the tapered element oscillating microbalance (TEOM)[29] was investigated for use in providing continuous mass measurements of nanoparticles generated artificially in an exposure chamber. The TEOM principle (developed initially for measuring the mass of particles in space) involves the use of a small filter which is located on the tip of a tapered glass tube which forms part of an oscillation microbalance (see schematic diagram in Figure 3.)

The oscillation frequency of the microbalance changes with the mass of particles collected on the filter. The devices are widely used to continuously monitor ambient levels of PM_{10} and $PM_{2.5}$ aerosols in national air quality networks and have proved to give reliable information on particle levels for compliance with national air quality directives. Consequently, with a mass detection limit of 0.01 µg, they were considered to possibly have adequate measurement precision (± 5 µg m^{-3} for 10 minute averaging times and ± 1.5 µg m^{-3} for 1 hour averaging times) for the measurement of nanoparticles in workplaces. In the Wake study, a range of polydisperse aerosols with number median mobility diameters in the nanometre range were dispersed into an exposure chamber in which were sited instruments for measuring number (SMPS) and surface area (DC LQ1) concentrations alongside the TEOM for mass. Despite being fitted with a pre-separator cutting at 1 µm, the TEOM overestimated the mass concentration of the finer aerosols when compared to the mass concentrations predicted by the SMPS, using the known density of aerosols generated. This was thought to be due to the presence of small numbers of large particles. In addition, and somewhat contradictorily, it was found that the standard Palflex filter used in the TEOM allows some nanoparticles to penetrate. The conclusion from this study was that careful consideration needs to be given to select a pre-separator for the TEOM to match the size

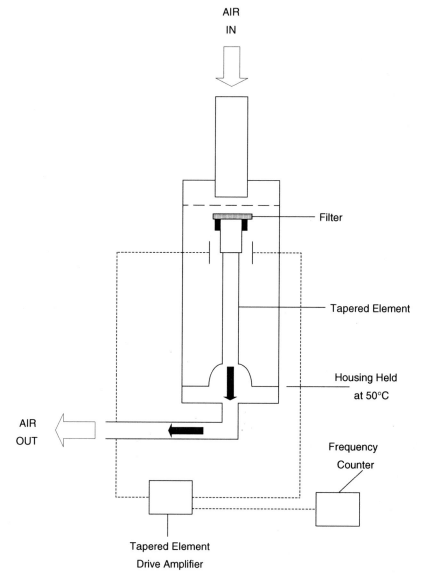

Figure 3 Principle of operation of Tapered Element Oscillating Microbalance
            (TEOM).

of nanoparticles being studied and to change the collection filter to one that has
high efficiency for nanoparticles. There is now a personal version of the TEOM
instrument (R & P PDM) developed for use in sampling respirable dust in coal
mines. The current version has a sensitivity about 10 times higher than that of
the standard version described above. However, with a change to a more
efficient filter, modification to a non-mining application and installation of a
suitable pre-selector it could be considered for use with nanoparticles.

3.3 Number Concentration

3.3.1 Condensation Particle Counters. The most widely used instrument for determining the number concentration of nanoparticles is the Condensation Particle Counter (CPC). This device exploits vapour condensation on nanometre size (and larger) particles in order to grow the particles to a size range that can be detected optically.

The convective cooling laminar flow CPC is the most widely used and is also commercially available from a number of manufacturers in models with different lower particle size cut-offs. A schematic diagram of a typical CPC is given in Figure 4.

Particle laden air is drawn into the instrument at constant flowrate, which is saturated using warm vapour (typically butanol, isopropanol or water). The saturated flow is then taken to a cool condenser tube in which the vapour is depleted onto the tube surface. However, as the flow cools, there will be regions in the flow where the vapour becomes supersaturated and condenses onto particles, which grow to large droplets. The detection limit at small particle diameters depends on vapour properties, operating temperatures (which determine the supersaturation), flows and geometries of the instrument. Devices using butanol are available with detection limits down to 3 nm, while isopropanol has successfully been used in portable instruments with a lower detection limit of 10 nm, and water is used in a commercially available instrument with a similar lower detection limit. The instrument can be used in connection with size classifying instruments such as a Differential Mobility Analyser (DMA), Scanning Mobility Particle Sizer (SMPS) or a diffusion battery to determine aerosol size distribution. Initially all CPCs were

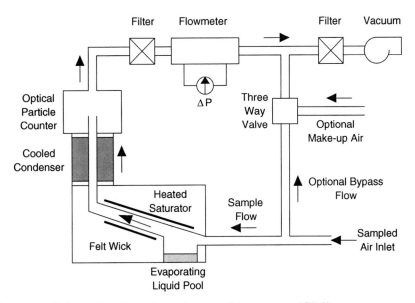

Figure 4 Schematic of condensation particle counter (CPC).

mains-powered and available from a number of manufacturers. However, at least one company supplies models of robust portable CPC – the TSI P-Trak, with lower detection size of 20 nm, and the TSI 33007, with lower detection size of 10 nm.

3.3.2 Electrometers. A second instrument type that is sensitive to nanoparticles is an electrometer. This instrument detects the charge carried by aerosol particles and therefore its use depends on knowing the charge on individual particles in an aerosol flow. Known charge distributions are possible to obtain using chargers or neutralizers with known characteristics. However, as charging efficiency is strongly a function of particle size, accurate information of the concentration of nanoparticles is difficult to obtain using an electrometer alone. An electrometer in series with a mobility analyser enables the determination of the size distribution of nanoparticles. In practice, the electrometer is often used to calibrate other instruments, especially CPCs, due to good detection efficiency at nanoparticle size range.

3.4 Surface Area Concentrations

Measurements of particle surface area have been possible for some time using the BET method.[30] However, it requires the collection of relatively large amounts of particles, and measurements are influenced by particle porosity (which may or may not be important) and collection/support substrate – particularly where the quantity of material analysed is small. The first instrument designed specifically to measure aerosol surface area was the epiphaniometer.[31] This device measures the Fuchs or active surface area of the aerosols by measuring the attachment rate of radioactive ions to the sampled particles. A particle pre-selector is required to exclude the unwanted large particles. Measurements of active surface area are generally insensitive to particle porosity. Unfortunately, the epiphaniometer is not well suited to widespread use in the workplace because it uses a radioactive source to provide the radioactive ions, for which a transport licence is required.

The same measurement principle is used in the diffusion charger aerosol surface-area monitors that have recently become available. These instruments measure the attachment rate of positive unipolar ions to particles, from which the aerosol active surface area is inferred.[32] A schematic diagram of the principle of a typical monitor is given in Figure 5.

The sampled aerosol passes through a weak plasma created by a corona discharge device where it mixes with the unipolar air ions produced by the corona. The air ions diffuse and attach to the exposed surface of the particles. The excess unattached ions are removed by a collecting electrode and the particles with attached charges are collected on a HEPA filter within a Faraday cup electrometer. The current produced by the charged particles is measured by a sensitive electrometer and related to the surface area of the sampled particles. Diffusion charging surface area monitors are available from a number of

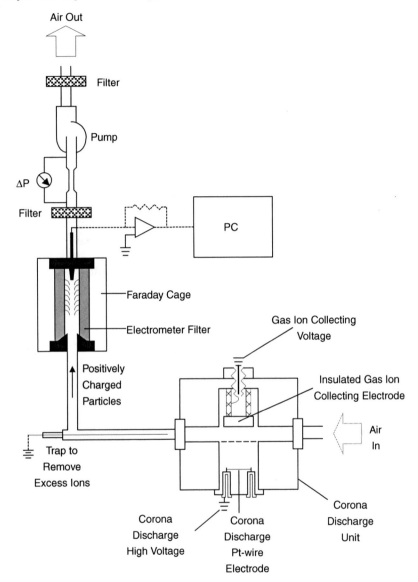

Figure 5 Schematic of diffusion charger surface area monitor.

companies (*e.g.* LQ1-DC from Matter Engineering, Switzerland, and DC2000CE from EcoChem, USA) and typically have quoted ranges of $0–2000 \ \mu m^2 \ cm^{-3}$ and sensitivities of $1 \ \mu m^2 \ cm^{-3}$. The latter is portable and battery-powered making it potentially more suitable for use in workplaces.

As yet, it is unknown how relevant active surface area is to health effects following inhalation exposure. Below approximately 100 nm active surface area has been found to correlate well with geometric surface area as measured by Scanning Mobility Particle Sizer (see Section 3.4.1) and with projected surface area as measured by Transmission Electron Microscopy.[33] However, above approximately 1 μm active surface area is a function of particle diameter, and so the relationship with actual particle surface area is lost.

A new device, the TSI Model 3550 Nanoparticle Surface Area Monitor,[34] uses a particular configuration of an aerosol charger to indicate the human lung-deposited surface area corresponding to the tracheobronchial and alveolar regions of the lung, rather than the total active surface area (*i.e.* Fuchs surface area) measured by the conventional diffusion charge instruments described above. The aerosol is drawn through a cyclone with a 1 μm cut point and then into the mixing chamber to mix with the ion stream. The voltage on the ion trap is altered such that it acts as a particle size selector to collect both the excess ions and particles that are not of a charge state (surface area size) corresponding to either the tracheobronchial or respirable aerosol fractions. The electric charges on the penetrating particles are then measured by the electrometer. The principle of this new device has been shown to provide a measurement that correlates well with deposited aerosol surface area in the lungs.[35] A portable battery-powered version of this instrument, the TSI Aerotrak 9000, has recently been launched.

3.5 Nanoparticle Size Distribution Measurement

3.5.1 Measuring Size Distribution using Particle Mobility Analysis. The most common instrument used for measuring size distributions of aerosols of nano-particles is the Scanning Mobility Particle Sizer (SMPS). The SMPS is capable of measuring aerosol size distribution in terms of particle mobility diameter from approximately 3 nm up to around 800 nm, although multiple instruments typically need to be operated in parallel to span this range. A schematic diagram is given in Figure 6.

It comprises an electrical aerosol analyser (EAA) to separate the particles according to the electrical mobility diameter followed by a CPC or an electrometer to count the particles. Particles enter a pre-selector with a cut-point at 1 μm into a region where they are then charged to Boltzmann equilibrium by passing them through a bipolar ion cloud formed from a radioactive source. They then pass through a well-defined electrostatic field in the EAA. Electro-static forces lead to charged particles moving between the electrodes, and particles with a specific mobility are sampled from a small outlet at the exit of the electrodes, form where they are passed to a CPC or electrometer for counting. By scanning the voltage between the electrodes, particles with electrical mobilities corresponding to a range of particle diameters can be counted sequentially, allowing the aerosol size distribution to be determined. In an alternative configuration, the voltage between the electrodes may be stepped rather than continuously scanned.

The sequential scanning or stepping of the voltage takes a significant time with the fastest conventional scan speeds being about 3 minutes, which is suitable provided that the process being monitored does not change within this timescale. Fast Mobility Particle Sizers have been developed using a unipolar

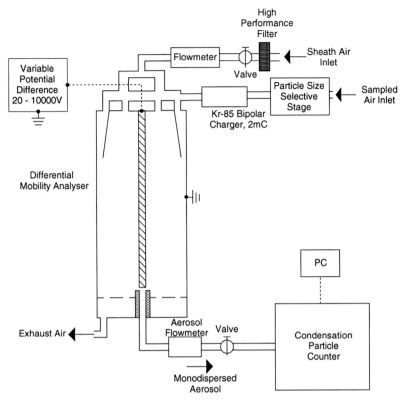

Figure 6 Schematic of Scanning Mobility Particle Sizer (SMPS).

particle charger, and a parallel array of electrometer-based sensors to count the size segregated particles. Measurements may be made with a time resolution of one second or less, and operation at ambient pressures reduces evaporation of volatile particles. The instruments are limited to measurements at relatively high aerosol number concentrations, although the lack of a radioactive source may make them a viable alternative to the SMPS in many workplaces. Research is currently being carried out to develop more compact, and therefore cheaper, aerosol mobility classifiers relying on particle migration across an opposing air flow,[36] and it is hoped that these will be available soon. Currently SMPS instruments are available from three main companies TSI (USA), Grimm (Austria) and Cambustion (UK).

The SMPS is limited in its widespread application in the workplace due to its size, expense, complexity of operation, the need for two or even three instruments operating in parallel to measure wide aerosol size distributions, and the use of a radioactive source to bring the aerosol to charge equilibrium.

3.5.2 Measuring Particle Size Distribution using Inertial Impaction. Cascade impactors are widely available in a number of configurations, allowing either personal or static sampling with a range of particle size cut points. Personal cascade impactors are available with cut-points of 250 nm and above, and thus are only able to provide very limited information on size distribution in the nanometre size range. Static cascade impactors are available with lower cut points in the nanometre size region, as well as low-pressure impactors or multi-orifice impactors.

There is number of low-pressure cascade impactors available and a recently developed one is the Dekati DLPI, which has 13 impactor stages with stage D_{50} cut-off sizes between 10 μm and 30 nm and a high-efficiency filter to collect the <30 nm particles. It can operate at flowrates of either 10 or 30 l min^{-1}, depending upon requirements. The stages are greased to prevent particle bounce causing contamination of the lower stages. The MOUDI Model 110 microorifice impactor consists of ten impactor stages, with D_{50} cut-off sizes between 18 μm and 56 nm; it also has a high-efficiency filter to collect the <56 nm particles. In the MOUDI, the collection substrates can be rotated relative to the impaction plates. This is to prevent the upper stages from becoming overloaded, resulting in coarse particle contamination of the lower stages, and enables the device to be operated for long periods of time. The MOUDI is operated at a flowrate of 30 l min^{-1}. The particles of aerodynamic diameter <56 nm can be further separated by feeding the output to an additional 3 stage Nano-MOUDI with D_{50} cut-off sizes of 32, 20 and 10 nm. Both impactors require high pressure, mains-powered pumps to provide the necessary flowrates and so are not suitable for personal sampling.

Determination of aerosol size distribution from cascade impactor data requires the application of data inversion routines. The simplest approach is to calculate cumulative mass concentration with particle diameter, and use the data to estimate the mass median aerodynamic diameter (MMAD) and the Geometric Standard deviation (GSD) of the size distribution. This approach assumes no losses between collection stages, ideal impactor behaviour, and a unimodal aerosol with a log-normal size distribution. Cascade impactors are usually used to measure the mass-weighted aerosol size distribution, and so assumptions of particle shape and density need to be made in order to estimate the number or surface-area weighted distribution. As these parameters are rarely quantified, great care needs to be taken in interpreting cascade impactor data in terms of aerosol number or surface area.

3.5.3 Electrical Low Pressure Impactor (ELPI) Measurements. The Electrical Low Pressure Impactor (ELPI) combines inertial collection with electrical particle detection to provide near-real-time aerosol size distributions for particles larger than 7 nm in diameter.[37] Aerosol particles are charged in a unipolar ion charger before being sampled by a low pressure cascade impactor discussed in Section 3.4.2. Each impactor stage is electrically isolated, and connected to a multi-channel electrometer, allowing a measurement of charge accumulation with time. As in the case of the diffusion charger (Section 3.3.3), particle charge is directly related to active surface area. Thus the integrated electrometer signal from all stages is directly related to aerosol active surface area.

The electrometer signal from a single stage is related to the active surface area of particles within a narrow range of aerodynamic diameters, allowing limited interpretation of the shape of sampled particles. If the particle charging efficiency as a function of aerodynamic diameter is known or can be assumed, real-time data from the ELPI can be interpreted in terms of the aerosol number-weighted size distribution. In practice, particle-charging efficiency is determined experimentally. Interpretation of measurements in terms of particle mass concentration or mass-weighted size distribution can also be carried out, although it requires the effective particle density as a function of size to be known.

As well as allowing on-line measurements of particle concentration and size distribution, aerosol samples collected by the ELPI are available for off-line analysis, including electron microscopy and chemical speciation. A diagram of the operating principle of the ELPI is shown in Figure 7.

3.5.4 Calculations of Nanoparticle Concentrations from Size Distribution Measurements. As well as providing information about the particle size

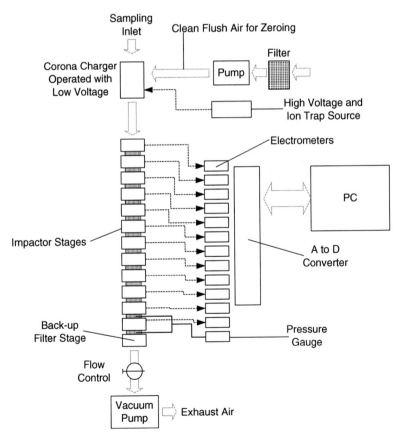

Figure 7 Diagram of operating principle of Electrical Low Pressure Impactor (ELPI).

characteristics of the aerosols in workplaces where nanoparticles are being produced or handled, size distribution measurements can be used to calculate integrated nanoparticle exposure levels. For example, number-weighted size distributions can be used simply to calculate number concentrations, or with the assumption that the particles are nearly spherical and that their physical diameters were equivalent to their mobility diameters (for SMPS, see below) or aerodynamic diameters (for LPI, see below), the aerosol surface concentration can be calculated. Similarly, with knowledge of particle density, the aerosol mass concentrations can be determined. However, the accuracy of these estimations is dependent upon the assumptions made about the physical characteristics of the particles. Ku and Maynard[33] showed that for monodisperse aerosol particles smaller than 100 nm, particle geometric surface areas calculated by SMPS size distributions agree to within ± 20% of those given by a diffusion charger surface area instrument. However, for larger particles, the relationship diverged with the DC instrument underestimating compared to the SMPS because of the change in response of the DC instrument. A similar relationship was found by Shi *et al.*[38] for polydisperse aerosols found in the ambient atmosphere. From comparative measurements at two outdoor sites they found good agreement between geometric surface area measurements using the epiphaniometer (see Section 3.3.3) and the SMPS. It is therefore reasonable to suggest that, provided that suitable pre-selectors are used with DC instruments, reliable measurements of geometric surface area can be obtained.

3.6   Particle Sampling Techniques for Characterisation

Determination of the physical and chemical properties of airborne nanoparticles relevant to their potential effect on human health is often required. Parameters such as particle size, shape, surface area, composition, crystallinity, solubility and bio-persistence provide the basic information for the exposure and toxicological evaluation of new nanomaterials. The surface coating on the particles and their electrical charge will also have a significant impact on their state of agglomeration, which will in turn influence their physical behaviour and subsequent biological responses.

The main analytical techniques routinely available for determining the particle size, shape and composition are high resolution electron microscopy combined with X-ray microanalysis and electron diffraction. Both scanning electron microscopy (SEM) and transmission electron microscopy (TEM) require samples of particles that are uniform in deposit and have minimal particle overlap. This rules out collection by impaction where particles are concentrated in small regions below the impaction jets, and for nanoparticles the filtration efficiency of most suitable filters is too low. For the SEM, particles down to typically 20 nm in diameter can be sampled directly onto SEM supports using electrostatic precipitation. Point-to-plane electrostatic precipitators combine a charging and deposition field by using a sharp corona needle as one electrode, and a planar collection surface as the second electrode. Sampling efficiency approaching 100% can be achieved for particles larger than 20 nm. For smaller particles, rapidly decreasing charging efficiency leads to a lower sampling efficiency. Deposits from electrostatic precipitators are generally uniform across the collection substrate, enabling discrete particle analysis in the SEM. A number of electrostatic precipitators is available from instrument manufacturers.

For the TEM, it is generally preferable to sample directly onto a TEM support grid, thus avoiding a secondary sample preparation stage. Thermal precipitation is the most suitable collection mechanism as it relies on aerosol particles migrating from a hot region to a cold region, and is particularly effective for particles between 1 nm and 100 nm in diameter. Thermal precipitation can be used to sample aerosols at ambient temperatures by establishing a temperature gradient above the collection surface, and passing the aerosol across the surface. A number of suitable designs have been published[39,40] and they can be built by a reasonable laboratory workshop.

3.7 Do Nanotubes Require Special Techniques?

Single-walled carbon nanotubes (SWCNT) essentially comprise a single layer of carbon atoms arranged in cylindrical structures of diameter about 1.5 nm and length up to about 1 mm. Carbon nanotubes may also form as multiple concentric tubes of diameters significantly greater than SWCNTs. The extreme aspect ratio of individual nanotubes, together with their potentially low solubility in the lungs may lead to toxic mechanisms analogous to those observed with other fibrous particles such as asbestos (see Chapter 5 for details). The question may be asked therefore if they should be considered, for exposure measurement purposes, like asbestos fibres and be analysed by counting under the TEM.

However, unlike asbestos, SWCNTs are very rarely found as single particles. They are generally produced as convoluted bundles of nanotubes (nanoropes) of diameter from 20 to 50 nm and then form complex clumps and agglomerates, of size between 100 μm and 1 mm, with other nanoropes and other carbonaceous and catalyst materials that are present. Laboratory and field studies by Maynard *et al.*[41] have shown that it is extremely difficult to break these clumps and generate aerosols of nanotubes. Normal procedures of transferring SWCNT powder from production vessel to storage bucket and then tipping into a second bucket showed no increase in nanoparticles numbers. It was only by using a single component vortex shaker fluidised bed, operating at over 50% agitation that any significant increase in particle numbers were produced. It is very doubtful that SWCNTs will be handled with such great force and so the likelihood of large numbers of discrete SWCNT particles being found in the workplace is low. However, for certain applications, manufacturers are currently trying to prevent nanoparticles from agglomerating by using some form of surface coating or other techniques. In addition, there is no information on the size distributions of particles released from the cutting, sanding or abrading of products that incorporate nanotubes bound into the matrix of the material (composites, tyres, *etc.*). Therefore at the moment there is no reason to suggest that nanotubes should be treated like asbestos fibres for exposure assessments, but it would be wise when monitoring levels of nanoparticles in workplaces where carbon nanotubes are being produced or handled to investigate samples collected for TEM analysis for discrete nanotubes. Finally, a careful watch should be kept on developments in nanotube production, and knowledge shared of any evidence of discrete airborne nanotubes found in workplace air.

3.8   Sampling Strategy Issues

Until it has been agreed which is (are) the most appropriate metric(s) for assessing exposure to nanoparticles in relation to potential adverse effects, it is recommended that a range of instrumentation be used to provide full characterisation of the aerosols in workplaces where nanoparticles are being produced, handled or used to make new materials. This results in large numbers of instruments that are not conducive to the normal personal sampling procedures that are required to assess personal exposure for compliance with any exposure limit or for epidemiological purposes.

However, new instruments are continuously being developed and there are small portable instruments for particle number concentrations (TSI P-Trak), particle surface area concentrations (EcoChem DC2000CE) and health-related surface area concentrations (TSI Aerotrak 9000). Whilst these instruments are not yet truly personal, they are compact enough to be carried from location to location in the workplace and to be sited close to the worker at each location. Currently however, these instruments do not provide enough information for full characterisation of the workplace and so static instruments such as the SMPS, ELPI and thermal/electrostatic precipitators for collecting particles for characterisation should be included. Care should be taken in siting these static samplers as aerosol characteristics can change with distance from source, leading to spatial and temporal variation of nanoaerosol mass and number concentration. This is especially true for hot processes leading to particle nucleation from vapour that will often lead to variations in emission rate and concentration over time.

To improve the comparability of exposure data, the accepted practice of giving personal exposure as an eight-hour-shift value should also be observed in the case of nanoaerosols. In consequence, wherever possible, exposure measurement results concerning shorter measurement intervals should be converted into shift data by time-weighted recalculation. In all cases, where short-term exposure itself is the target of investigations, the time base of measurements needs to be documented. A time base of 15 minutes for short-term exposure measurements is recommended as it is generally used in occupational hygiene.

Selection of the most appropriate sampling location or locations is a key factor for a reliable interpretation of data in view of personal exposure. This requires identification of all the potential nanoaerosol-emitting sources in the workplace and an understanding of the ventilation system in the workplace to determine the potential for cross contamination. This could be a significant problem for nanoparticles as they will remain airborne for considerable periods of time and be easily dispersed by the air currents in the workplace. For single sources, the relationship between aerosol emission and work activities should be clear, enabling the reliable assignment of exposure levels to be made.

However, unless the workplace is operating under clean room conditions or has high efficiency filters on the inlet air through well defined inlets, outdoor sources of nanoaerosols (*e.g.* vehicle exhausts, other industrial activities, power stations, *etc.*) will penetrate indoors and result in overestimation of the levels of nanoparticles emitted from the process under investigation. This will inevitably lead to an overestimation of the worker exposure to nanoparticles derived from that process. The simplest way to overcome this problem is to carry out simultaneous measurement of background concentrations using a duplicate set of monitoring equipment to monitor outside the workplace, and to subtract the outdoor levels from those measured inside the workplace. However, this can be expensive and assumes that the ambient particles do not change during transport into the workplace.

Alternatively, differences in composition between nanoparticles generated in the workplace and those combustion particles in the outdoor air can be used for discrimination purposes. If the composition of the engineered nanoparticles is known, and the constituent elements are not likely to be found in outdoor air, then the proportion of the engineered particles in the total particle field counted by TEM and analysed by X-ray microanalysis can be determined. This ratio can then be used to calculate the surface area concentration of the engineered nanoparticles in the total surface area concentration values for all airborne nanoparticles detected. The accuracy of this approach will obviously depend upon the engineered nanoparticles having at least one detectable element that is not present in outdoor aerosols. This proposition has not yet been fully tested.

4 Review of Reported Measurements of Exposure to Nanoparticles

4.1 Introduction

There are very few published data on human exposure measurements of airborne nanoparticles in the workplace in which specific attention has been paid to identify and measure the particles of size fraction < 100 nm. As mentioned in Section 1.2, ultrafine particles are produced as unwanted by-products in processes such as welding, soldering, thermal spraying and coating, and in the exhaust gases from combustion engines, but until very recently they were sampled for occupational exposure purposes as respirable or even inhalable aerosols and analysed gravimetrically (welding) or chemically (diesel exhaust). In addition, aerosol exposure measurements have also been carried out in the workplaces of existing manufacturing processes, such as carbon black, fumed silica, ultrafine titanium dioxide, fine nickel powder and precious metal blacks, following the current guidance given by national governments and international standards bodies. However, this guidance does not currently include specific methods for assessing exposure to nanoparticles.

4.2   Measurements of Nanoparticle Exposures in Existing Industries

Two main studies have been reported in which levels of nanoparticle exposures in existing workplaces have been assessed using instrumentation relevant for nanoparticles. In 2000, Wake[42] carried out a review of industries based in the UK to identify those that produce, handle or generate nanoparticles or nano-powders, and to identify the processes that give rise to airborne nanoparticles. This was followed by short visits to a selection of those companies identified and measurements of nanoparticle exposures in their relevant workplaces. The main parameters measured were: a) particle number concentrations using the TSI portable CPC, the P-Trak or its predecessor the TSI Portacount, with minimum particle size detected of 20 nm, and b) particle number size distributions using the TSI SMPS set to give particle size distributions in the range 16.5 to 805 nm. Measurements were made as close as possible to the person operating the process, giving an indication of the levels available for exposure. At each workplace, ambient particle number concentrations were measured outdoors during the sampling shift, to determine the level of ultrafine particles entering the workplace. In addition, details of the processes being carried out in the workplaces visited were recorded and other observations made to enable future comparisons to be made with similar data and therefore to gain maximum benefit in future risk assessments. Summaries of the number concentrations and size distributions are given in Table 2, for production of nanoparticles in existing industries, and Table 3, for existing industries where nanoparticles are unwanted by-products.

In each table, the workplace levels can be compared with the outdoor ambient levels, with the idea of subtracting the latter from the former to calculate the levels of nanoparticles actually produced by the process itself. However, this is not necessarily valid as it is clear that for some of the workplaces, ambient number concentrations are higher than those found in the workplaces themselves. An example of this is in the carbon black production where very high levels of particles outside the factory in the loading bay area were thought to be due to diesel exhaust fumes from delivery lorries and smoke from the flaring off at a large oil refinery close to the factory. The median diameters of the particles close to the bagging plant were generally similar to those outdoors indicating no evidence of primary or aggregated carbon black particles. However, occasionally high median diameters were recorded during the bagging of fluffy and pelletised carbon. This can be seen in Figure 8.

This experience was confirmed by Kuhlbusch et al.,[43] who, in a large study of airborne particles in the carbon black industry, found that carbon black

Table 2 Summary of airborne nanoparticle levels in existing industries in UK (Wake[42]).

| Industry | Activity | No. conc. range Particles $cm^{-3} \times 10^3$ | | Number-based particle size distributions | | | |
| | | | | Ambient | | Workplace | |
		Ambient	Workplace	No. med. dia. (nm)	O'_g	No. med. dia. (nm)	O'_g
Carbon black[a]	Bagging	694–3836	4–50	44	3.2	51 to 400	2.4
Nickel powder	Bagging	3–16	4–212	23	1.9	49	3.3
Precious metal blacks	Sieving	19–62	23–71	NM	NM	NM	NM
Titanium dioxide	Bagging	10–58	4–17	NM	NM	NM	NM
Zinc refining	Sintering	20–23	12–24	NM	NM	NM	NM
Zinc refining	Casting	20–23	56–100	503	5.3	70	2.2

[a]Total number concentrations by SMPS, NM = Not measured.

Table 3 Summary of airborne nanoparticle levels as by-products in existing industries in UK (Wake[42]).

| Industry | Activity | No. conc. range Particles $cm^{-3} \times 10^3$ | | Number-based particle size distributions | | | |
| | | | | Ambient | | Workplace | |
		Ambient	Workplace	No. med. dia. (nm)	O'_g	No. med. dia. (nm)	O'_g
Plasma coating[a]	Wire coating	2–8	3–905	41	2.2	587	1.3
Galvanising[a]	Galvanising	15–37	10–683	64	2.0	99	2.1
Steel foundry	Moulding	13–73	118–>500	46	1.9	66	2.0
Welding	MIG	10–19	117–>500	53	2.1	179	2.2
Plastic welding[a]	Welding	1–5	111–3766	31	2.0	37	1.7
Hand soldering	Tinning	2–11	12–>500	41	2.0	72	2.3

[a]Total number concentrations by SMPS.

particles released by bag filling activities had a size distribution starting at aerodynamic diameters of 400 nm with modes at 1 μm and 8 μm. They also suggested that any nanoparticles measured in the bag filling areas were attributed to diesel exhaust emissions from forklift trucks. A similar situation was found at the titanium dioxide factory where power was being transferred during the finishing and bagging plants. In this case, however, Wake suggested that the recirculating local exhaust system operating in the factory may have been responsible for cleaning the air inside the factory.

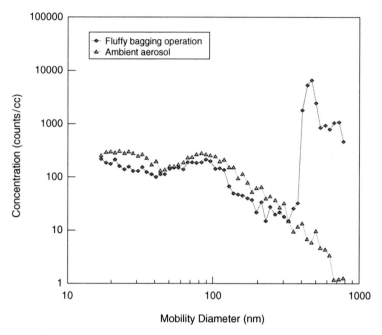

Figure 8 Number-based size distributions at carbon black bagging plant (Wake[42]).

Table 4 Summary of airborne ultrafine particle levels in some workplaces in Germany (Mohlmann[25]).

Process	Number concentration in size range 14 to 673 nm Particles $cm^{-3} \times 10^3$	Mode of size distribution nm
Silica smelting	100	280–520
Metal grinding	130	17–170
Soft soldering	400	36–64
Plasma cutting	500	120–180
Bakery	640	32–109
Airport terminal	700	< 45
Hard soldering	54–3500	33–126
Welding	100–40000	40–600

The measurements carried out on levels of unwanted by-product nanoparticles have been replicated in a long-term ongoing study of ultrafine particle levels in German industry workplaces.[25] In this study, factories were chosen for investigation on the basis that they include processes with high temperatures or combustion. Number-based size distribution measurements were made using a TSI SMPS operated to measure particles in the size range 14 to 675 nm. In addition, a Berner-type low pressure cascade impactor was used to estimate the mass concentrations of the ultrafine particles experienced. A summary of the results is given in Table 4.

Replicated measurements are available for two processes. For hand soldering and welding, the number concentrations and the size distributions are similar for each study. Similar number concentrations for MIG welding measured using the SMPS of 100×10^3 were reported by Brouwer *et al.*[26] in a study designed to explore measurement methods for assessing worker exposure to airborne nanoparticles.

4.3 Measurements of Nanoparticle Exposures in New Nanotechnology Processes

I have been unable to identify any published studies that have reported investigations into worker exposure to nanoparticles from new nanotechnology processes. As portable instruments become available for measuring levels of airborne nanoparticles, measurements will be made in workplaces with new nanotechnology processes (manufacture or use) to assess the adequacy of the control measures and to assess personal exposures. One such brief study was reported by Shakesheff.[44] Using the TSI P-Trak portable CPC he found that particle number concentrations inside the nanopowder manufacturing plant were lower than those outdoors and particle levels in the production rig exhaust were an order of magnitude lower, indicating effective performance of the powder control systems. However, the most significant measurements will be made in the national and international research programmes designed to provide data for risk assessment and regulation.

In the meantime, it is informative to discuss some research carried out in 2002 by Maynard *et al.*[41] into potential inhalation and dermal exposure from handling unrefined single-walled carbon nanotube (SWCNT) material. Measurements were made at four facilities where very low density material, comprising carbon nanotubes, nanometre-sized catalyst metal particles and other forms of elemental carbon, were produced. Inhalation and dermal exposure were measured during the removal of the material from production vessels and handling of the material prior to processing. To overcome the confounding factor of variable background levels of ambient particles (mainly carbon from vehicle exhausts), all work was carried out outside in a specially built clean air enclosure that reduced ambient levels to below 50 particles cm^{-3}. Particle levels inside the enclosure during the various handling operations carried out were monitored using the TSI Model 3007 CPC with lower detection limit of 10 nm and a GRIMM series 1.100 optical particle counter with monitoring range 300 nm to 15 µm. In addition, personal and area samples were taken during the SWCNT handling procedures using open face filter holders. The authors concluded that the results showed that handling the nanotube material did not lead to an increase in the number concentration of fine particles, as measured by the CPC. They went on to suggest that released particles tend to be larger than 1 µm in diameter and any increases of nanoparticles measured were considered to be due to particle release from the vacuum cleaner motor or to releases of small nanotube clusters from the vacuum cleaner used to clean up the powder. The personal air samples gave dust concentrations of between 0.7 and 53 μg m^{-3}, but small numbers of large clumps of material were found on the filters that would have dominated the mass measurement, as the samplers were not equipped with particle pre-selectors.

Measurements of dermal exposure were made using cotton gloves fitted on top of the protective gloves. Glove deposits of SWCNT during handling were estimated at between 0.2 and 6 mg per hand and the authors suggested that dermal contact could be a significant exposure route if relevant protective measures are not followed.

5   Discussion

As with all new and emerging technologies, the development of reliable techniques for assessing and controlling exposure to nanoparticles in the workplace will always be working from a position of insufficient knowledge until the suitability of current controls are assessed, the emissions rates of nanoparticles from those processes are determined and the exposures of the workforce to those nanoparticles are characterised. Together with information on the toxicity of the nanoparticles to human health, these parameters form the basis of the risk assessment process that informs legislation on its production, sale and use, allows the setting of appropriate occupational exposure limits and leads to guidance on the choice of suitable control procedures.

The area is moving fast and instrument manufacturers are currently developing new devices that they hope will become the mainstay of future nanoparticle exposure assessments. Besides the recently introduced TSI Aerotrak 9000 (see Section 3.4), the author has heard of a number of developments in the pipeline, including personal CPCs, small portable DC surface area monitors, small, portable instruments that provide particle number size distributions (similar to the information provided by the SMPS) and small, portable particle mass monitors. In addition, there are many other long-term developments including a possible portable device that should be able to discriminate between engineered and combustion nanoaerosols. So, assuming that international agreement can be obtained about which metric or metrics is the most appropriate to use as the basis of exposure assessment for inhalation of airborne nanoparticles, then the future looks promising that a suitable sampling methodology will be available. The choice of sampler or monitor depends upon the role for which it is to be used and a device for exposure assessment may be different from that used to determine sources and to assess the efficiency of control systems. For example, for inhalable and respirable aerosols personal filter-based samplers are used for exposure assessment, whilst direct-reading dust monitors are used for assessing control systems and finding sources. For nanoparticles, we are in the fortunate position of having collaboration on health and safety issues at the early stages between researchers, producers and users and governments and so it is to be hoped that any decisions that are made are accepted by all parties. To help this process, the Nanoparticle Occupational Safety and Health Consortium[45] – a group of 14 members including: nano-particle manufacturers and users and government bodies, led by DuPont – has produced a set of requirements that it wants in an instrument for daily monitoring of R&D and manufacturing facilities. This was presented to instrument manufacturers at the 2006 AIHCE conference and it is hoped that a suitable device will soon become available.

Current guidance in the UK for controlling worker exposure to substances harmful to health is based around a control banding scheme called COSHH Essentials produced by HSE. This employs a decision-tree approach in which the answers to a number of questions about the process and the material used lead the user to the most appropriate control solution. The questions include a description of the process, the chemical to be used, its human health hazard (toxicity), the amount used and, for powders, its dustiness. The usefulness of this approach is that with a small number of essential pieces of information about the process and the material, a potentially suitable control system can be chosen. For this reason, it is planned to extend the control banding approach to nanomaterials. One of the important parameters to measure for the material is its "dustiness". This is defined as the propensity for dust to be released into the air when the material is handled and two bench-top reference methods are given in the European Standard EN 15051.[46] These methods have been deve-loped and validated using a range of different materials from different indus-trial sectors including titanium dioxide, carbon black, baker's flour, atapulgite, barium sulfate, *etc.* and even including chicken layers pellets. Whilst one material (carbon black) does contain nanostructured particles, it has not been determined whether these methods will be suitable for nanopowders. The author's laboratory has recently carried out some preliminary tests using the rotating drum method with three different nanopowders and the indications are that dustiness values are very low and the metric for assessment may have to be changed. Some early work with SWCNTs was carried out by Maynard *et al.*[41] and they found that the materials were difficult to disperse and required the considerable energy deployed in a two component fluidised bed with fluidisa-tion of the beads provided by a standard laboratory vortex shaker to produce any significant particle release. Further work on developing suitable methods for classifying nanopowders in terms of their dustiness should be carried out, although care should be taken to ensure that the methods developed give classification results that relate to the relative amounts of dust that is released from materials in real handling processes.

It is clear that there is currently a dearth of information about worker exposure to new engineered nanoparticles. It is expected that this situation will improve when the results from research currently being carried out reach the public domain. There are two main EU-funded projects that include exposure assessment that will report in the next few years. NANOSAFE2[24] is investi-gating the safe production and use of nanomaterials in three main workplaces, whilst NANOSH, a new project due to start in autumn 2006, intends to obtain worker exposure data in 16 different workplaces throughout the EU, including some in university research laboratories. The data from these will be combined into a common database and so it is vitally important that the settings of equipment used (such as the particle size range of the SMPS) are agreed beforehand so that the results can be combined. There is also much work being carried out by national governments throughout the world and strong efforts are being made to agree a common sampling strategy and reporting requirements. The target is for as large a database as possible to provide the necessary information for the risk assessment process.

References

1. The Royal Society and The Royal Academy of Engineering, *Nanoscience and Nanotechnologies: Opportunities and Uncertainties*, The Royal Society, London, 2004.
2. E. Hood, *Environ. Health Perspect.*, 2004, **112**, A741.
3. R.J. Aitken, K.S. Creely and C.L. Tran, *Nanoparticles: an occupational hygiene review*, HSE Research Report number 274, 2004.
4. A.D. Maynard and E.D. Kuempel, *J. Nanoparticle Res.*, 2005, **7**(6), 587.
5. J.S. Tsuji, A.D. Maynard, P.C. Howard, J.T. James, C.-W. Lam, D.B. Warheit and A.B. Santamaria, *Toxicol. Sci.*, 2006, **89**(1), 42.
6. A.D. Maynard, *Nano Today*, 2006, **1**(2).
7. M.C. Roco, *J. Nanoparticle Res.*, 2003, **5**(3–4), 181.
8. BSI, *Vocabulary: Nanoparticles*, PAS 71, 2005.
9. ISO, *Workplace Atmospheres – Ultrafine, nanoparticle and nano-structured aerosols – Inhalation exposure characterization and assessment*, ISO/TR 27628, 2007.
10. G. Oberdörster, E. Oberdörster and J. Oberdörster, *Env. Health Perspec.*, 2005, **113**(7), 823.
11. ICRP, *Human respiratory tract model for radiological protection*, ICRP publication 66, 1994.
12. S. Tinkle, in *Proceedings of First International Symposium on Health Implications of Nanomaterials, 12–14 October 2004, Buxton UK*, 2005, p. 47.
13. ISO, *Air Quality – Particle size fraction definitions for health-related sampling*, IS 7708, 1995.
14. HSE, *Asbestos: The analyst's guide for sampling, analysis and clearance procedures*, HSG 248, 2005.
15. CEN, *Workplace atmospheres – guidelines for measurement of airborne micro-organisms and endotoxin*, BS EN 13098, 2001.
16. G. Oberdörster, *Phil. Trans. Roy. Soc. London Series A*, 2000, **358**(1775), 2719.
17. K. Donaldson, V. Stone, P.S. Gilmour, D.M. Brown and W. MacNee, *Phil. Trans. R. Soc. London Series A*, 2000, **358**(1775), 2741.

18. D.M. Brown, M.R. Wilson, W. MacNee, V. Stone and K. Donaldson, *Toxicol. Applied Pharmacology*, 2001, **175**(3), 191.

19. C.L. Tran, D. Buchanan, R.T. Cullen, A. Searl, A.D. Jones and K. Donaldson, *Inh. Toxicol.*, 2000, **12**(12), 1113.

20. A. Nemmar, P.H.M Hoet, B. Vanquickenborne, D. Dinsdale, M. Thomeer, M.F. Hoylaerts, H. Vanbilloen, L. Mortelmans and B. Nemery, *Circulation*, 2002, **105**, 411.

21. G. Oberdörster, Z. Sharp, V. Atudorei, A. Elder, R. Gelein, W. Kreyling W and C. Cox, *Inh. Toxicol.*, 2004, **16**(6–7), 437.

22. N.A. Fuchs, *The Mechanics of Aerosols*, Pergamon, Oxford, 1964.

23. R.M. Harrison, J.-P. Shi, S. Xi, A. Khan, D. Mark, R. Kinnersley and J. Yin, *Phil. Trans. R. Soc. London Series A*, 2000, **358**(1775), 2567.

24. NANOSAFE2 project: see website http://www.nanosafe.org.

25. C. Möhlmann, *Gefahrstoffe-Reinhaltung der Luft*, 2005, **65**(11/12), 469.

26. D.H. Brouwer, J.H.J. Gijsbers and M.W.M. Lurvink MWM, *Ann. Occup. Hyg.*, 2004, **48**, 439.

27. G.R. Cass, L.A. Hughes, P. Bhave, M.J. Kleeman, J.O. Allen and L.G. Salmon, *Phil. Trans. R. Soc. London Series A*, 2000, **358**(1775), 2581.

28. D. Wake, *An investigation into the relationship between mass, number and surface area and the influence of particle composition and morphology, for instruments measuring laboratory simulated workplace aerosols containing ultrafine and nanoparticles*, 2006, HSE Research Report Number 513.

29. E. Rupprecht, M. Meyer and H. Patashnick, *J. Aerosol Sci.*, 1992, **23** (Suppl 1), S635.

30. S. Brunauer, P.H. Emmett and Edward Teller, *J. Amer. Chem. Soc.*, 1938, **60**(2), 309.

31. U. Baltensberger, H.W. Gäggeler and D.T. Jost, *J. Aerosol Sci.*, 1988, **19**(7), 931.

32. A. Keller, M. Fierz, K. Siegmann, H.C. Siegmann and A. Filippov, *J. Vac. Sci. Technol. A*, 2001, **19**(1), 1.

33. B.K. Ku and A.D. Maynard, *J. Aerosol Sci.*, 2005, **36**(9), 1108.

34. TSI Inc., Model 3550 Nanoparticle Surface Area Monitor, details from www.tsi.com.

35. W.E. Wilson, in *Proceedings of the 2004 Air and Waste Management Association Conference*, 2004.

36. R.C. Flagan, *Aerosol Sci. Technol.*, 2004, **38**(9), 890.

37. J. Keskinen, K. Pietarinen and M. Lehtimäki *et al.*, *J. Aerosol Sci.*, 1992, **23**(4), 353.

38. J.-P. Shi, R.M. Harrison and D.E. Evans, *Atmos. Environ.*, 2001, **35**(35), 6193.

39. A.D. Maynard, *Aerosol. Sci. Technol.*, 1995, **23**, 521.

40. S. Plitzko, in *Proceedings of BIA workshop Ultrafine Aerosols at the Workplace*, 2003, 127.

41. A.D. Maynard, P.A. Baron, M. Foley, A.A. Shvedova, E.R. Kisin and V. Castranova, *J. Toxicol. Environ. Health*, 2004, Part A, **67**, 87.

42. D. Wake, *Ultrafine particles in the workplace*, 2001, HSL Report number ECO/00/18.
43. T.A.J. Kuhlbusch, S. Neumann and H. Fissan, *J. Occup. Environ. Health*, 2004, **1**, 660.
44. A.J. Shakesheff, 2005, Private communication.
45. NOSH project: see website www.dupont.com.
46. CEN, *Workplace atmospheres: measurement of the dustiness of materials*, EN 15051, 2006.

Toxicological Properties of Nanoparticles and Nanotubes

KEN DONALDSON AND VICKI STONE

1 Introduction

The toxicology of manufactured/engineered nanoparticles (NPs) is of increasing concern as nanotechnology continues to develop and manufactured NPs are progressively developed.[1] The existing toxicology database on nanoparticles is almost entirely based on combustion-derived nanoparticles in environmental air. This research led to the "ultrafine hypothesis",[2,3] which emphasised the combustion-derived nanoparticle (CDNP) component of particulate matter (PM) as a key component mediating adverse health effects. The toxicological mechanisms are centred around the ability of particles to cause oxidative stress and inflammation and translocate from the site of deposition.[4] This review introduces environmental nanoparticles and their mechanisms as a basic paradigm, and then moves on to discuss toxicology of engineered nanoparticles.

2 Environmental Air Pollution Particles

2.1 Effects of Environmental Particles

The adverse health effects of air pollution have been recognised throughout much of recorded time (Table 1). Burning of fossil fuels in towns and cities, where there is little mixing of air, during periods of cold weather has been associated with the generation of smogs consisting mainly of sulfur dioxide and particles. Particles or particulate matter (PM) represent a part of the air pollution cocktail present in ambient air, which also comprises gases such as ozone, nitrogen dioxide, *etc.* Particulate material in ambient air is measured as the mass of particles collected using the PM_{10} or $PM_{2.5}$ sampling conventions.[5] The adverse health effects of PM_{10} are seen at the levels that pertain in the UK

Issues in Environmental Science and Technology, No. 24
Nanotechnology: Consequences for Human Health and the Environment
© The Royal Society of Chemistry, 2007

Table 1   Adverse health effects due to PM after reference 7.

Mortality from cardiovascular and respiratory causes
Admission to hospital for cardiovascular causes
Exacerbations of asthma in pre-existing asthmatics
Symptoms and use of asthma medication in asthmatics
Exacerbations of chronic obstructive pulmonary disease
Lung function decrease
Lung cancer

and other countries today and there is often no threshold. In other words there is a background of ill health being caused by PM that increases when the ambient particle cloud increases in concentration and decreases when the mass of particles in the air decreases.[6]

These adverse health effects of air pollution have been measured in hundreds of studies and there is good coherence between the acute effects seen in time series and panel studies, and the chronic effects seen in environmental studies.

Within a short time-lag of one or two days following an increase in PM there are increases in the following:

(i) all-cause mortality
(ii) attacks of asthma and increased usage of asthma medication
(iii) deaths in chronic obstructive pulmonary disease (COPD) patients
(iv) exacerbations of COPD
(v) deaths and hospitalisations for cardiovascular disease.[8]

The adverse cardiovascular effects associated with increases in PM are well documented. Panel studies have documented associations between elevated levels of particles and:

(i) onset of myocardial infarction[9]
(ii) increased heart rate[10]
(iii) decreased heart rate variability.[11]

Studies exposing subjects to concentrated airborne particles (CAPs) have also shown increased lung inflammation[12] and altered brachial artery diameter in relation to increased exposure.[13] A recent epidemiological study in the USA measured carotid intima-media thickness (CIMT) in life,[14] a measure of atherosclerosis, and demonstrated an association between atherosclerosis and ambient air pollution level. Living in an area with a 10 μg m^{-3} higher level of $PM_{2.5}$ was associated with a CIMT increase of 5.9% (95% confidence interval, 1–11%); an even larger effect, 15.7%, was seen in older women.

2.2   Nanoparticles as the Drivers of Environment Particle Effects

Environmental PM is a complex mixture of particle types that depend on season, time of day, siting of sampler, *etc*. Combustion-derived nanoparticles, however,

Table 2 Common components of PM and comments on their origin, nature and likely toxic potency.

PM_{10} component	Comment	Toxic potency
Combustion-derived nanoparticles	Nanoparticles containing metals and organic volatiles; derived from combustion *e.g.* vehicle exhaust particles	High
Sodium/magnesium compounds	Derived from sea spray	Low
Sulfate	Predominantly ammonium sulfate	Low
Nitrate	Predominantly ammonium nitrate	Low
Calcium/potassium compounds and insoluble minerals	Derived from the earth's crust *e.g.* clay	Low
Biologically derived materials	*e.g.* Endotoxin	High

represent a major toxicologically important component (see below). The common components of PM are shown in Table 2, along with an indication of their toxic potency. Sulfates are found to be virtually non-toxic in experimental toxicology studies (see review[15]) but do show a relationship with adverse effects in some epidemiological studies, *e.g.* reference 16; this apparent anomaly may be explained by a correlation between sulfates and some more potent component of the air pollution mix which is actually driving the adverse effect, such as fine particles.

Nanoparticle number concentrations across three European cities were recently found[17] to range from 15 000 to 18 000 particles per cm^3, and from 10 000 to 50 000 particles per cm^3 in a busy London street.[18] In a study on US highways, exposure in a vehicle travelling in busy traffic was reported to be 200 to 560 × 10^3 particles per cm^3 (predominantly nanoparticles)[19,20], whilst indoor cooking, vacuuming and burning wax candles produce nanoparticles of soot.[21] Nanoparticles also arise from combustion of domestic gas and in one study three gas rings produced around 50 000 particles per cm^3, which underwent rapid aggregation, with increases in particle size and decrease in apparent number within a few minutes.[22] Secondary nanoparticles may also arise from environmental chemistry, *e.g.* nitrates, but these are unlikely to be as toxicologically potent as combustion-derived nanoparticles (see below). The mechanisms of the adverse effects of combustion-derived nanoparticles has been extensively reviewed by the authors.[4,23,24]

The cellular molecular mechanisms involved in generation of local lung inflammation by combustion-derived NPs are shown in Figure 1.

In this paradigm surfaces, organics and metals can all contribute to inflammation via the production of free radicals. Diesel exhaust particles (DEPs) represent a good exemplar of combustion-derived NPs. Diesel exhaust particles cause inflammation in rat[25,26] and human lungs[27] following short-term, high-level exposure. Oxidative stress is demonstrable as an increased level of 8 OH dG, the oxidative adduct of hydroxyl radical, in the lungs of rats following exposure and in cells in culture treated with DEP.[28,29] The component of DEP responsible for

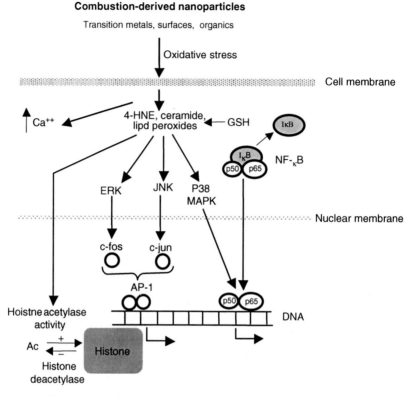

Figure 1 Hypothetical series of events leading from combustion-derived NPs such as diesel soot interactions with lung cells leading to inflammatory gene expression.

the oxidative stress and subsequent pro-inflammatory signalling is principally the organic fraction,[30–33] although transition metals may also be involved, especially for other nanoparticles such as welding fume.[34] The oxidative stress then causes activation of signalling pathways for pro-inflammatory gene expression, including MAPK[30,35–37] and NF-κB activation[30,38] and histone acetylation that favours pro-inflammatory gene expression.[39] Activation of these pathways culminates in transcription of a number of pro-inflammatory genes such as IL-8 in epithelial cells treated *in vitro*[40] and in human lungs exposed by inhalation.[41] TNFα has been reported to be increased in macrophages exposed to DEP *in vitro*[42] and IL-6 is released by primed human bronchial epithelial cells exposed to DEP.[43]

3 Could Cardiovascular Effects of PM be Due to CDNP?

The well-documented cardiovascular endpoints that are affected by PM[44,45] could be mediated through the effects of CDNP. In a number of studies using a mixture of PM, CAPs and model NP studies, evidence is accumulating that

NPs cause inflammation that could adversely affect the cardiovascular system. There is evidence of systemic inflammation following increases in PM, as shown by elevated C-reactive protein, blood leukocytes, platelets, fibrinogen and increased plasma viscosity (reviewed in reference 46). Atherosclerosis is a key process in atherothrombosis, the main cause of cardiovascular morbidity and mortality.[47] Atherosclerosis is an inflammatory process, initiated via endo-thelial injury and producing systemic markers of inflammation that are risk factors for myocardial and cerebral infarction.[47–49] Repeated exposure to PM_{10} may, by increasing systemic inflammation, exacerbate the vascular inflamma-tion of atherosclerosis and promote plaque development or rupture. Experi-mental studies with animal models susceptible to atherosclerosis confirm the ability of particle exposure to enhance atherosclerosis.[50,51] These effects are summarised on the left side of Figure 2.

Particles in the lungs could also, either directly by entering the airway walls or via inflammation, stimulate nerve endings that could alter the normal neural regulation of the heart leading to fatal dysrhythmia.[52,53] In support of this hypothesis, studies in humans and animals have shown changes in the heart rate and heart rate variability in response to particle exposures.[10,11,52,54]

Particles themselves may also enter the blood and directly impact on the clotting system or on atherothrombotic processes in the plaques.[55,56] There is evidence that particles can translocate from the lungs to the blood in animal models[57,58] and affect platelets[59] and thrombus formation.[59] Humans exposed to diesel exhaust nanoparticles showed endothelial dysfunction in the forearm, suggesting that local deposition of CDNPs in the lungs has a critical effect on the endothelial bed throughout the body.[60] These effects are summarised on the right side of Figure 2.

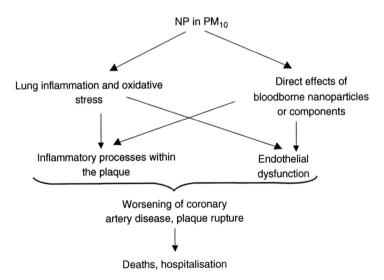

Figure 2 Hypothetical pathways from inflammation to local pulmonary and cardiovascular effects.

4 Is the Environmental Nanoparticle Paradigm Applicable to Engineered NPs?

If the environmental paradigm was generalisable to new engineered NPs we would anticipate that new NPs would cause oxidative stress and inflammation in the lungs, affect the cardiovascular system to enhance atherothrombosis and also modulate heart rate variability. They could also translocate to the brain with unknown consequences (see later). So far studies with new NPs have been almost completely confined to toxicity studies on cells and to animal studies for carbon nanotubes. Little is known about their potential to affect the cardiovascular system. As described below, however, there is a wide range of composition and shape of nanoparticles and these would be expected to impart considerable differences in toxicity.

4.1 The Nature of Newer Manufactured Nanoparticles

There are a number of bulk-produced nanoparticles that have been used for decades in industry and these are typified by TiO_2 and carbon black (CB). In addition to these, the Emerging Nanotechnologies project, conducted by The Woodrow Wilson International Center for Scholars, along with the Pew Charitable Trusts, has identified, as at April 2006, over 200 consumer products into which newer nanotechnology, especially nanoparticles, is incorporated (see http://www.nanotechproject.org/index.php?id=44). The reasons for the rapid expansion in industrial use of this technology are its unique properties due to small size and large reactive surface area. Nanoparticles come in a wide variety of shapes, sizes and chemical compositions. In addition to the spherical shapes observed for particles such as TiO_2, shape varieties also include carbon nanotubes, nanowhiskers and nanofibres. Engineered nanoparticles vary considerably in their size and composition and so would be anticipated to vary in toxicity. Nanotubes and nanowires can range from less than 100 nm diameter to tens of μm in length. The variety of chemical composition ranges from substances considered traditionally to be relatively inert (*e.g.* carbon and gold) to substances associated with significant toxicity (*e.g.* cadmium and other heavy metals). Since the NP size imparts heightened reactivity compared to the "inert" materials, it is interesting to consider the impact of small size on toxic materials, especially since reactivity might relate to toxicity.

4.2 Carbon Black and TiO_2

Carbon black and TiO_2 along with other NPs, have been studied for some time with regard to their pro-inflammatory effects. There have been few or no studies on the effects of these materials on the cardiovascular system. Nano-size carbon black has been intensively studied with regard to the issue of low toxicity dust and the confounding effect of rat lung overload (extensively reviewed in references 61–64). Rat lung overload describes the phenomenon whereby large surface area doses of particles of low toxicity cause inflammation and failed clearance in rat lungs, culminating in fibrosis and cancer.[65–68] In view of the very high surface area per unit volume of NPs and the identification of surface area as the driver[69–71], attention has focused on the nanoparticulate form of the nuisance dusts. These would be anticipated to produce lung overload at lower mass lung burdens than those seen with the larger particles. This was indeed found to be the case.[71,72] However, Gilmour *et al.*[73] showed that even at low, non-overload exposures to nanoparticulate carbon black (NPCB), there was a pro-inflammatory effect not seen with the larger CB particles.

Instillation studies have also shown that the nanoparticulate form of CB and TiO_2 produce more inflammation than an equal mass of larger, yet respirable, particles[74,75] of the same material and across a range of nanoparticles of nuisance dust the surface area was found to be the driver of the inflammation.[76]

The molecular mechanism of the increased inflammatory effects of nanoparticle carbon black have demonstrated that they generate ROS in cell-free systems[77,78,79] and cause alterations in calcium signaling[80,81] in exposed cells. Oxidative stress from the NPCB can also activate the Epidermal Growth Factor (EGF) receptor[82] and redox-responsive transcription factors such as NF-κB[81] and AP-1[83] leading to the transcription of pro-inflammatory cytokines and lipid mediators.[79,81]

4.3 Nanoparticles and the Brain

Oberdörster demonstrated that inhaled carbon NPs could gain access to the rat brain via the olfactory nerves in the roof of the nasal cavity.[84] Another study showed that only 14 nm and not 95 nm carbon black caused alterations in brain inflammatory parameters after inhalation.[85] Further research on the meaning and consequences of such findings are warranted given the importance of neural control of the cardiovascular response.

4.4 New Engineered NPs and the Cardiovascular System

Only one study at present has examined the role of new NPs on the clotting system *in vitro*. Radomski *et al.*[86] studied the effects of multi-wall and single-wall nanotubes, C_{60} fullerenes and mixed carbon black nanoparticles on human platelet aggregation *in vitro* and rat vascular thrombosis *in vivo*. Standard urban particulate matter was used as a control. Nanotubes and carbon black particles, but not C_{60}, stimulated platelet aggregation and the same ranking was seen in ability to affect the rate of vascular thrombosis in rat carotid arteries; urban dust had low activity in these assays. Thus, there are differences between different carbon nanoparticles to activate platelets and enhance vascular thrombosis.

4.5 Carbon Nanotubes

Carbon nanotubes (CNTs) are long sheets of graphite rolled in the form of a tube, which can be a few nm thick (single-walled; SWCNT) up to a few hundred nm thick (multi-walled; MWCNT). The needle-like structure implies that a paradigm related to fibres such as asbestos might be appropriate in considering their toxicity. Length greater than 20 μm, thinness and biopersistence are the descriptors of the potential pathogenicity of a conventional fibre.[87] Biopersistence is an important determinant of mineral fibre and synthetic vitreous fibre pathogenicity. Long biopersistent fibres are the biologically effective dose that drives pathogenic effects[87] whilst non-biopersistent fibres undergo dissolution processes that can be enhanced at the acid pH of 5.0 existing inside macrophage phagolysosomes.[88] Long biosoluble fibres undergo leaching of key structural molecules leading to breakage into short fibres that are readily phagocytosed by the macrophages.[87,89]

Using the analogy of asbestos fibres the biopersistence of nanotubes would be likely to play an important role in pathogenicity. In one study[90] both unground and ground nanotubes 0.7 and 5.9 μm long respectively were assessed for biopersistence and the longer, unground nanotubes were more biopersistent than the short ones. This is consistent with the greater biopersistence of long fibres seen in studies with asbestos and other mineral fibres although these "long" nanotubes were much shorter than those mineral fibres defined as "long", which are in the region of 20 μm and greater.[87] A 20 μm diameter rat macrophage is able to enclose and transport fibres less than its own diameter from the lungs,[87] and the length-dependent inhibition of clearance seen with 5.9 μm long nanotubes is thus rather unexpected. It may be that the well-documented tendency for nanotubes to form bundles and wires[91] is important in controlling their clearance.

Nanotubes have been used in a number of rat lung instillation studies. However, the high dose and dose-rate raise questions about physiological relevance; no study has addressed the role of length by comparing long (> ∼20 μm) with short (< 10 μm) nanotubes. Lam and co-workers demonstrated that a single intratracheal instillation in mice with three different types of SWCNT (size not stated) caused dose-dependent granulomas and interstitial inflammation.[92] Quartz and carbon NPs at equal mass dose were used as controls and the authors concluded that, on an equal mass basis, SWCNTs in the lungs were more toxic than carbon black and even quartz. The acute lung toxicity of intratracheally instilled SWCNTs that were 1.4 nm wide by > 1 μm long, and that contained appreciable amounts of Ni, Co and amorphous carbon, was determined by Warheit *et al.*[93] SWCNTs produced a transient inflammation and a non dose-dependant accumulation of multifocal granulomas. The significance of these granulomas was, however, questioned as their formation was possibly due to the instillation of a bolus of agglomerated nanotubes.[93]

Shvedova *et al.* studied mice exposed to SWCNTs[94] that were 99.7% elemental carbon and 0.23% iron by weight. Two morphologies were seen, compact aggregates and dispersed structures, and these caused two distinct responses: the dense SWCNT aggregates were associated with foci of granulomatous inflammation and discrete granulomas with hypertrophic epithelial cells while exposure to the dispersed SWCNTs caused diffuse interstitial fibrosis. Interestingly the SWCNTs caused a dose-dependent decrease in lung glutathione, and an increase in 4-hydroxynonenal, indicative of oxidative stress.

Carbon nanotubes have been tested in a range of different cell types *in vitro* to assess their potential toxicity. Treatment of human keratinocytes have shown that both SWCNTs and MWCNTs are capable of being internalised and causing cellular toxicity.[95,96] In a study with alveolar macrophages, SWCNTs were more cytotoxic than MWCNTs after exposure at equal mass dose.[97] Human T cells exposed to oxidised MWCNTs were killed in a time- and dose-dependent manner, with apoptosis being involved,[98] as found also in kidney cells exposed to SWCNTs.[99] Manna *et al.* demonstrated dose-dependent oxidative stress and NF-κB activation in human keratinocytes along with IκB depletion and MAPK phosphorylation.[100]

The repetitive structure of crystals of long asbestos has been implicated in causing aggregation of the Epidermal Growth Factor (EGF) receptor and initiating signalling cascades[101] relevant to asbestos disease and the EGF-R receptor is activated by NPCB.[82] The ordered graphene structure of a long carbon nanotube might also have some ability to cause receptor aggregation. The idea that the graphene structure could be important is supported by the finding that the more the surface of SWCNTs is derivatised/functionalised the lesser is its *in vitro* toxicity.[102] However, it should be noted that the ability to kill cells *in vitro*, at high dose, is not necessarily reflective of the ability to cause disease.

At the moment it is difficult to draw general conclusions on CNT toxicity because of the paucity of data and their variability. They can very in length, and composition, including metal contamination. Carbon nanotubes are often kinked and tangled into aggregates of varying size and shape. This kind of variability is found between and within samples. All of these factors could impact on toxicity. Our own experience is that the more rigid CNTs disperse more efficiently than the tangled ones. However the tangled CNTs are more easily taken up by cells in culture and could thus be more readily cleared from the lungs by macrophages. Clearly much more work is needed to define the factors that drive CNT toxicity and so characterisation of individual CNT samples used in studies is imperative.

4.6 Fullerenes

Buckminster fullerene or C_{60} is a compact, cage-like molecule comprising 60 carbon atoms. C_{60} can be viewed as a nanoparticle and has received some toxicological attention. The basic graphene structure of fullerene can be functionalised in various ways to change the physical properties of the fullerene, for instance making it more dispersible/soluble.[103] Carboxylated fullerenes are slightly less toxic than the native C_{60} and hydroxylated fullerenes are virtually non-toxic to human dermal fibroblasts.[103] The toxicity of the native C_{60} appeared to be due to the ability to generate superoxide anions.[103] Isakovic *et al.* showed a similar effect with the native C_{60} being 30 times more toxic on a mass basis than a soluble hydroxylated C_{60} in a range of human tumour cell lines. The authors concluded through the use of antioxidants, and by identifying different types of cell death with the two fullerenes, that unmodified C_{60} had strong pro-oxidant capacity responsible for the rapid necrotic cell death whilst polyhydroxylated C_{60} exerted mainly antioxidant/cytoprotective effects and produced modest apoptosis that was independent of oxidative stress.[104] Unmodified C_{60} was not toxic to guinea pig alveolar macrophages *in vitro* in another study, whilst nanotubes were toxic at the same mass dose.[97] More research into more sophisticated endpoints than toxicity *in vitro* is needed to clarify the potential toxicity of C_{60}. Paradoxically, several studies have suggested that C_{60} is an antioxidant[105-108] and antinitrosating[105] agent. Clearly more research is warranted to resolve the issue as to the role of the C_{60} surface chemistry derivatisation *versus* pristine C_{60} and the issue of antioxidant potential.

4.7   Quantum Dots

Quantum dots (QDs), also known as nanocrystals, are a unique class of semiconductor because of their 2–10 nanometers size. At such small particle sizes the electrons are strongly confined leading to emission of light. Emission wavelengths vary in response to size with smaller dots yielding "bluer" wavelengths and larger dots giving "redder" wavelengths. To achieve these properties they are composed of an element each from different periodic table groups.[109] Although there is little potential for inhalation exposure to QDs, they have considerable possibility for imaging and diagnosis *in vivo* and so there is potential for exposure of a number of target organs reached from the blood.

Quantum dots can vary considerably in their elemental composition and elements such as cadmium have considerable toxic potential. Cadmium Qds[110] have been found to be toxic to cells in cultures and this effect was inhibited by the thiol N-acetyl cysteine, suggesting a role for oxidative stress. Other studies demonstrate low toxicity in some models of cellular toxicity.[111,112] Likewise animal studies have demonstrated a lack of acute toxicity.[111,113] Clearly more research is needed on QDs to understand fully their potential toxicity.

4.8   Other Nanoparticles

The toxicity of a range of nanoparticles with similar singlet particle size (average 20–50 nm) was tested[114] and revealed striking differences in their cytotoxicity, ranging from a virtually non-toxic rutile TiO_2 NP to highly toxic silver nanoparticles; many NPs however were intermediate in toxicity towards the RAW mouse macrophage cell line used in the study. In another study, a panel of NPs of similar size (30–45 nm) was tested in liver cells.[115] CDNPs were the most toxic, followed by Ag, whilst Mo, Al, Fe and TiO_2 had similar low toxicity. The mechanism of toxicity was assessed for Ag particles and a dramatic increase in intracellular fluorescence, indicative of oxidative stress, was seen in cells loaded with DCF-DH and exposed to Ag NPs. This was accompanied by depletion of GSH and cell death, confirming that oxidative stress may have been caused by loss of mitochondrial membrane potential.[115] The toxicity of copper was tested in the form of copper ions, NP copper and micron-sized copper. The greatest toxicity to the liver, kidney and spleen was seen with the nano-copper following oral dosing compared to the same mass dose of micron-sized copper.[116]

In a study by Dick *et al.* a range of NPs of different composition were compared for ability to induce ROS production *in vitro* versus ability to cause lung inflammation in rats.[117] There was a good correlation between the two, suggesting that the pro-inflammatory effects of these different NPs were related to their ability to generate oxidative stress.

5 Conclusion

A paradigm has evolved arising from experience with environmental, combustion-derived nanoparticles exemplified by diesel soot. In this paradigm oxidative stress and inflammation were identified as key processes in the local effects in the lungs. In addition inflammatory effects and blood translocation could explain adverse cardiovascular effects seen in epidemiology studies with air pollution particles. Support for this contention came from a number of studies using model NPs and CDNPs, where adverse cardiovascular effects such as clotting plaque development and endothelial dysfunction were enhanced after NP exposures in a number of different models. In parallel with these studies, an increasing number of toxicology studies used bulk NPs such as TiO_2 and carbon black and these identified a key role for the large surface area and its ability to produce oxidative stress.[79,118,119] The important question that arises is whether the same paradigm can be used for new engineered nanoparticles and nanotubes. In the limited studies so far published, engineered nanoparticles such as the carbon nanotubes are also reported to induce oxidative stress, cell death and inflammation. However, it is important to point out that there are variations in the degree of the adverse effects shown by NPs in various models and so not all are likely to show the same toxic potency. This is to be anticipated given our understanding that the total toxicity of any particle sample is the complex sum of the surface reactivity, times the surface area, plus releasable toxic moities, plus shape, all modified by biopersistence. There is a strong likelihood that all of these can vary considerably and so the total toxicity is very likely to vary between particle types.

The other variable is exposure. Even hazardous particles require some exposure and very little is known about exposure to the newer engineered NPs – these data are urgently needed. Taken together the data suggest that, for some of the new engineered NPs, sufficient exposure could lead to oxidative stress and a stimulation of the inflammatory response, and that this would be linked to adverse health effects similar to those seen with CDNPs (Figure 3). More research is needed on translocation and the role that this plays in adverse cardiovascular and potential neurological effects of the new engineered NPs.

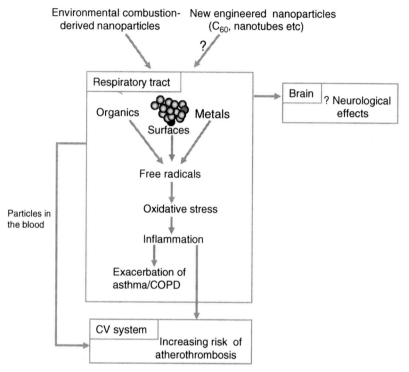

Figure 3   Summary of the paradigm for the harmful effects of combustion-derived NPs, raising the question of whether new engineered nanoparticles – a very diverse group of materials – could have the same paradigm.

References

1. Royal Society and Royal Academy of Engineering: Nanoscience and nanotechnologies: opportunities and uncertainties. *The Royal Society*, 2004.
2. A. Seaton, W. MacNee, K. Donaldson and D. Godden, Particulate air-pollution and acute health-effects, *Lancet*, 1995, **345**, 176–178.
3. M.J. Utell and M.W. Frampton, Acute health effects of ambient air pollution: the ultrafine particle hypothesis, *Journal of Aerosol Medicine*, 2000, **13**, 355–359.
4. K. Donaldson, L. Tran, L. Jimenez, R. Duffin, D.E. Newby, N. Mills, W. MacNee and V. Stone, Combustion-derived nanoparticles: A review of their toxicology following inhalation exposure, *I. Part. Fibre Toxicol.*, 2005, **2**(1), 10.
5. Quality of Urban Air Review Group, *Airborne particulate matter in the United Kingdom: third report of the Quality of Urban Air Review Group.* 1996. *Quality of Urban Air Review Group.*

6. B. Brunekreef and S.T. Holgate, Air pollution and health, *Lancet*, 2002, **360**(9341), 1233–1242.

7. C.A. Pope, III and D.W. Dockery, Epidemiology of particle effects 197, *Air Pollution and Health*. ed. S.T. Holgate, J.M. Samet, H.S. Koren and R.L. Maynard, Academic Press, San Diego, 1999, pp. 673–705.

8. C.A. Pope and D.W. Dockery, Epidemiology of particle effects, , in *Air Pollution and Health*, ed. S.T. Holgate, J.M. Samet, H.S. Koren and R.L. Maynard, Academic Press, San Diego, 1999, pp. 673–705.

9. A. Peters, D.W. Dockery, J.E. Muller and M.A. Mittleman, Incresed particulate air pollution and the triggering of myocardial infarction, *Circulation*, 2001, **103**(23), 2810–2815.

10. H. Gong Jr., W.S. Linn, S.L. Terrell, K.W. Clark, M.D. Geller, K.R. Anderson, W.E. Cascio and C. Sioutas, Altered heart-rate variability in asthmatic and healthy volunteers exposed to concentrated ambient coarse particles, *Inhal. Toxicol.*, 2004, **16**(6-7), 335–343.

11. R.B. Devlin, A.J. Ghio, H. Kehrl, G. Sanders and W. Cascio, Elderly humans exposed to concentrated air pollution particles have decreased heart rate variability, *Eur. Respir. J. Suppl*, 2003, **40**, 76s–80s.

12. A.J. Ghio, C. Kim and R.B. Devlin, Concentrated ambient air particles induce mild pulmonary inflammation in healthy human volunteers, *Am. J. Respir. Crit. Care Med.*, 2000, **162**(3 Pt 1), 981–988.

13. R.D. Brook, J.R. Brook, B. Urch, R. Vincent, S. Rajagopalan and F. Silverman, Inhalation of fine particulate air pollution and ozone causes acute arterial vasoconstriction in healthy adults, *Circulation*, 2002, **105**(13), 1534–1536.

14. N. Kunzli, M. Jerrett, W.J. Mack, B. Beckerman, L. LaBree, F. Gilliland, D. Thomas, J. Peters and H.N. Hodis, Ambient air pollution and atherosclerosis in Los Angeles, *Environ. Health Perspect.*, 2005, **113**(2), 201–206.

15. R.B. Schlesinger and F. Cassee, Atmospheric secondary inorganic particulate matter: the toxicological perspective as a basis for health effects risk assessment, *Inhal. Toxicol.*, 2003, **15**(3), 197–235.

16. C.A. Pope, M.J. Thun, M.M. Namboodiri, D.W. Dockery, J.S. Evans, F.E. Speizer and C.W. Heath, Jr., Particulate air pollution as a predictor of mortality in a prospective study of U.S. adults, *American Journal of Respiratory & Critical Care Medicine*, 1995, **151**(3 Pt 1), 669–674.

17. J.J. de Hartog, G. Hoek, A. Mirme, T. Tuch, G.P. Kos, H.M. ten Brink, B. Brunekreef, J. Cyrys, J. Heinrich, M. Pitz, T. Lanki, M. Vallius, J. Pekkanen and W.G. Kreyling, Relationship between different size classes of particulate matter and meteorology in three European cities, *J. Environ. Monit.*, 2005, **7**(4), 302–310.

18. Air Quality Expert Group: Particulate matter in the United Kingdom. *Published by the Department of Environment, Food and Rural Affairs, London*, 2005.

19. A. Elder, R. Gelein, J. Finkelstein, R. Phipps, M. Frampton, M. Utell, D.B. Kittelson, W.F. Watts, P. Hopke, C.H. Jeong, E. Kim, W. Liu,

W. Zhao, L. Zhuo, R. Vincent, P. Kumarathasan and G. Oberdorster, On-road exposure to highway aerosols. 2. Exposures of aged, compromised rats, *Inhal. Toxicol.*, 2004, **16**(Suppl 1), 41–53.

20. D.B. Kittelson, W.F. Watts, J.P. Johnson, M.L. Remerowki, E.E. Ische, G. Oberdorster, R.M. Gelein, A. Elder, P.K. Hopke, E. Kim, W. Zhao, L. Zhou and C.H. Jeong, On-road exposure to highway aerosols. 1. Aerosol and gas measurements, *Inhal. Toxicol.*, 2004, **16**(Suppl 1), 31–39.

21. A. Afshari, U. Matson and L.E. Ekberg, Characterization of indoor sources of fine and ultrafine particles: a study conducted in a full-scale chamber, *Indoor. Air*, 2005, **15**(2), 141–150.

22. M. Dennekamp, S. Howarth, C.A. Dick, J.W. Cherrie, K. Donaldson and A. Seaton, Ultrafine particles and nitrogen oxides generated by gas and electric cooking, *Occup. Environ. Med.*, 2001, **58**(8), 511–516.

23. K. Donaldson, L.A. Jimenez, I. Rahman, S.P. Faux, W. MacNee, P.S. Gilmour, P.J. Borm, R.P.F. Schins, T. Shi, V. Stone, Repiratory health effects of ambient air pollution particles: Role of reactive species. in *Oxygen/nitrogen radicals: lung injury and disease*, ed. V. Vallyathan, X. Shi and V. Castranova, Volk 187 in *Lung Biology in Health and Disease Exec*, ed. C. Lenfant, Marcel Dekker, New York, 2004.

24. K. Donaldson, V. Stone, P.J. Borm, L.A. Jimenez, P.S. Gilmour, R.P. Schins, A.M. Knaapen, I. Rahman, S.P. Faux, D.M. Brown and W. MacNee, Oxidative stress and calcium signaling in the adverse effects of enviromental particle (PM(10)), *Free Radic. Biol. Med.*, 2003, **34**(11), 1369–1382.

25. Y. Miyabara, R. Yanagisawa, N. Shimojo, H. Takano, H.B. Lim, T. Ichinose and M. Sagai, Murine strain differences in airway inflammation caused by diesel exhaust particles, *European Respiratory Journal*, 1998, **11**(2), 291–298.

26. Y. Tsurudome, T. Hirano, H. Yamato, I. Tanaka, M. Sagai, H. Hirano, N. Nagata, H. Itoh and H. Kasai, Changes in levels of 8-hydroxyguanine in DNA, its repair and OGGI mRNA in rat lungs after intratracheal administration of diesel exhaust particles, *Carcinogenesis*, 1999, **20**(8), 1573–1576.

27. C. Nordenhall, J. Pourazar, A. Blomberg, J.O. Levin, T. Sandstrom and E. Adelroth, Airway inflammation following exposure to diesel exhaust: a study of time kinetics using induced sputum, *Eur. Respir. J.*, 2000, **15**(6), 1046–1051.

28. T. Ichinose, Y. Yajima, M. Nagashima, S. Takenoshita, Y. Nagamachi and M. Sagai, Lung carcinogenesis and formation of 8-hydroxy-deoxyguanosine in mice by diesel exhaust particles, *Carcinogenesis*, 1997, **18**(1), 185–192.

29. T. Arimoto, T. Yoshikawa, H. Takano and M. Kohno, Generation of reactive oxygen species and 8-hydroxy-2'-deoxyguanosine formation from diesel exhaust particle components in L1210 cells, *Jpn. J. Pharmacol.*, 1999, **80**(1), 49–54.

30. V. Bonvallot, A. Baeza-Squiban, A. Baulig, S. Brulant, S. Boland, F. Muzeau, R. Barouki and F. Marano, Organic compounds from diesel

exhaust particles elicit a proinflammatory response in human airway epithelial cells and induce cytochrome p450 1A1 expression, *Am. J. Respir. Cell. Mol. Biol.*, 2001, **25**(4), 515–521.

31. S. Hirano, A. Furuyama, E. Koike and T. Kobayashi, Oxidative-stress potency of organic extracts of diesel exhaust and urban fine particles in rat heart microvessel endothelial cells, *Toxicology*, 2003, **187**(2–3), 161–170.

32. N. Li, M.I. Venkatesan, A. Miguel, R. Kaplan, C. Gujuluva, J. Alam and A. Nel, Induction of heme oxygenase-I expression in macrophages by diesel exhaust particle chemicals and quinones via the antioxidant-responsive element, *J. Immunol.*, 2000, **165**(6), 3393–3401.

33. A.E. Nel, D. az-Sanchez and N. Li, The role of particulate pollutants in pulmonary inflammation and asthma: evidence for the involvement of organic chemicals and oxidative stress 37, *Curr. Opin. Pulm. Med.*, 2001, **7**(1), 20–26.

34. J.D. McNeilly, L.A. Jimenez, M.F. Clay, W. MacNee, A. Howe, M.R. Heal, I.J. Beverland and K. Donaldson, Soluble transition metals in welding fumes cause inflammation via activation of NF-kappaB and AP-11, *Toxicol. Lett.*, 2005, **158**(2), 152–157.

35. F. Marano, S. Boland, V. Bonvallot, A. Baulig and A. Baeza-Squiban, Human airway epithelial cells in culture for studying the molecular mechanisms of the inflammatory response triggered by diesel exhaust particles, *Cell. Biol. Toxicol.*, 2002, **18**(5), 315–320.

36. T.S. Hiura, M.P. Kaszubowski, N. Li and A.E. Nel, Chemicals in diesel exhaust particles generate reactive oxygen radicals and induce apoptosis in macrophages, *J. Immunol.*, 1999, **163**(10), 5582–5591.

37. S. Hashimoto, Y. Gon, I. Takeshita, K. Matsumoto, I. Jibiki, H. Takizawa, S. Kudoh and T. Horie, Diesel exhaust particles activate p38 MAP kinase to produce interleukin 8 and RANTES by human bronchial epithelial cells and N-acetylcysteine attenuates p38 MAP kinase activation, *Am. J. Respir. Crit. Care Med.*, 2000, **161**(1), 280–285.

38. H. Takizawa, T. Ohtoshi, S. Kawasaki, T. Kohyama, M. Desaki, T. Kasama, K. Kobayashi, K. Nakahara, K. Yamamoto, K. Matsushima and S. Kudoh, Diesel exhaust particles induce NF-kappa B activation in human bronchial epithelial cells in vitro: importance in cytokine transcription, *J. Immunol.*, 1999, **162**(8), 4705–4711.

39. P.S. Gilmour, I. Rahman, K. Donaldson and W. MacNee, Histone acetylation regulates epithelial IL-8 release mediated by oxidative stress from environmental particles, *Am. J. Physiol. Lung. Cell. Mol. Physiol.*, 2003, **284**(3), L533–L540.

40. N. Terada, N. Hamano, K.I. Maesako, K. Hiruma, G. Hohki, K. Suzuki, K. lshikawa and A. Konno, Diesel exhaust particulates upregulate histamine receptor mRNA and increase histamine-induced IL-8 and GM-CSF production in nasal epithelial cells and endothelial cells [see comments], *Clin. Exp. Allergy*, 1999, **29**(1), 52–59.

41. S.S. Salvi, C. Nordenhall, A. Blomberg, B. Rudell, J. Pourazar, F.J. Kelly, S. Wilson, T. Sandstrom, S.T. Holgate and A.J. Frew, Acute exposure to

diesel exhaust increases IL-8 and GRO-alpha production in healthy human airways, *Am. J. Respir. Crit. Care. Med.*, 2000, **161**(2 Pt 1), 550–557.

42. H.M. Yang, J.C. Ma, V. Castranova and J.H. Ma, Effects of diesel exhaust particles on the release of interleukin-1 and tumor necrosis factor-alpha from rat alveolar macrophages, *Experimental Lung Research*, 1997, **23**, 269–284.

43. P.A. Steerenberg, J.J. Zonnenberg, J.A. Dormans, P.T. Joon, I.M. Wouters, L. vanBree, P.J. Scheepers and H. VanLoveren, Diesel exhaust particles induced release of interleukin 6 and 8 by (primed) human bronchial epithelial cells (BEAS 2B) in vitro, *Experimental Lung Research*, 1998, **24**, 85–100.

44. R.D. Brook, B. Franklin, W. Cascio, Y. Hong, G. Howard, M. Lipsett, R. Luepker, M. Mittleman, J. Samet, S.C. Smith, Jr. and I. Tager, Air pollution and cardiovascular disease: a statement for healthcare professionals from the Expert Panel on Population and Prevention Science of the American Heart Association, *Circulation*, 2004, **109**(21), 2655–2671.

45. R.D. Brook, J.R. Brook and S. Rajagopalan, Air pollution: the "Heart" of the problem, *Curr. Hypertens. Rep.*, 2003, **5**(1), 32–39.

46. K. Donaldson, N. Mills, W. MacNee, S. Robinson and D. Newby, Role of inflammation in cardiopulmonary health effects of PM, 6, *Toxicol. Appl. Pharmacol.*, 2005, **207**(2 Suppl), 483–488.

47. J.F. Viles-Gonzalez, S.X. Anand, C. Valdiviezo, M.U. Zafar, R. Hutter, J. Sanz, T. Rius, M. Poon, V. Fuster and J.J. Badimon, Update in atherothrombotic disease, *Mt. Sinai. J. Med.*, 2004, **71**(3), 197–208.

48. P. Libby, P.M. Ridker and A. Maseri, Inflammation and atherosclerosis 79, *Circulation*, 2002, **105**(9), 1135–1143.

49. F. Van Lente, Markers of inflammation as predictors in cardiovascular disease, *Clin. Chim. Acta.*, 2000, **293**(1–2), 31–52.

50. T. Suwa, J.C. Hogg, K.B. Quinlan, A. Ohgami, R. Vincent and S.F. van EEDEN, Particulate air pollution induces progression of atherosclerosis, *J. Am. Coll. Cardiol.*, 2002, **39**(6), 935–942.

51. Q. Sun, A. Wang, X. Jin, A. Natanzon, D. Duquaine, R.D. Brook, J.G. Aguinaldo, Z.A. Fayad, V. Fuster, M. Lippmann, L.C. Chen and S. Rajagopalan, Long-term air pollution exposure and acceleration of atherosclerosis and vascular inflammation in an animal model, 1, *JAMA*, 2005, **294**(23), 3003–3010.

52. H.C. Routledge, S. Manney, R.M. Harrison, J.G. Ayres and J.N. Townend, Effect of inhaled sulphur dioxide and carbon particles on heart rate variability and markers of inflammation and coagulation in human subjects, 9, *Heart*, 2006, **92**(2), 220–227.

53. H.C. Routledge, J.G. Ayres and J.N. Townend, Why cardiologists should be interested in air pollution, *Heart*, 2003, **89**(12), 1383–1388.

54. L.C. Chen and J.S. Hwang, Effects of subchronic exposures to concentrated ambient particles (CAPs) in mice. IV. Characterization of acute

and chronic effects of ambient air fine particulate matter exposures on heart-rate variability, *Inhal. Toxicol.*, 2005, **17**(4–5), 209–216.

55. A. Nemmar, M.F. Hoylaerts, P.H. Hoet and B. Nemery, Possible mechanisms of the cardiovascular effects of inhaled particles: systemic translocation and prothrombotic effects, *Toxicol. Lett.*, 2004, **149**(1-3), 243–253.

56. K. Donaldson, N. Mills, W. MacNee, S. Robinson and D.E. Newby, Role of inflammation in cardiopulmonary health effects of PM, *Toxicol. Appl. Pharmacol.*, 2005.

57. W. Kreyling, M. Semmler, F. Erbe, P. Mayer, S. Takenaka, G. Oberdorster and A. Ziesenis, Minute translocation of inhlaed ultrafine insoluble iridium particles from lung epithelium to extrapulmonary tissues, *Ann. Occup. Hyg.*, 2002, **46**(Suppl 1), 223–226.

58. A. Nemmar, H. Vanbilloen, M.F. Hoylaerts, P.H. Hoet, A. Verbruggen and B. Nemery, Passage of intratracheally instilled ultrafine particles from the lung into the systemic circulation in hamster, *Am. J. Respir. Crit. Care Med.*, 2001, **164**(9), 1665–1668.

59. A. Nemmar, P.H. Hoet, D. Dinsdale, J. Vermylen, M.F. Hoylaerts and B. Nemery, Diesel exhaust particles in lung acutely enhance experimental peripheral thrombosis, *Circulation*, 2003, **107**(8), 1202–1208.

60. N.L. Mills, H. Tornqvist, S.D. Robinson, M. Gonzalez, K. Darnley, W. MacNee, N.A. Boon, K. Donaldson, A. Blomberg, T. Sandstrom and D.E. Newby, Diesel exhaust inhalation causes vascular dysfunction and impaired endogenous fibrinolysis, 2, *Circulation*, 2005, **112**(25), 3930–3936.

61. B. Bellmann, H. Muhle, O. Creutzenberg and R. Mermelstein, Irreversible pulmonary changes induced in rat lung by dust overload, *Environmental Health Perspectives.*, 1992, **97**, 189–191.

62. P.E. Morrow, Dust overloading of the lungs: update and appraisal, *Toxicol. Appl. Pharmacol.*, 1992, **113**(1), 1–12.

63. J.L. Mauderly, Y.S. Cheng and M.B. Snipes, Particle overload in toxicological studies - friend or foe, *Journal Of Aerosol Medicine-Deposition Clearance And Effects In The Lung*, 1990, **3**(Suppl 1), S169–S187.

64. P.E. Morrow, Possible mechanisms to explain dust overloading of the lungs, *Fundam. Appl. Toxicol.*, 1988, **10**(3), 369–384.

65. P.E. Morrow, J.K. Haseman, C.H. Hobbs, K.E. Driscoll, V. Vu and G. Oberdorster, The maximum tolerated dose for inhalation bioassays - toxicity vs overload, *Fundamental And Applied Toxicology*, 1996, **29**.

66. J.L. Mauderly, R.J. McCunney, Particle overload in the rat lung: implications for human risk assessment, *Inhalation Toxicology*, 1996, **8** (Supplement), p. 298.

67. J.L. Mauderly, Lung Overload:the dilemma and opportunities for resolution. Special issue. Particle overload in the rat lung and lung cancer:implications for risk assessment, *Inhal. Toxicol.*, 1996, **8** (Supplement), 1–28.

68. G. Oberdorster, Lung particle overload - implications for occupational exposures to particles, *Regulatory Toxicology And Pharmacology*, 1995, **21**.

69. C.L. Tran, D. Buchanan, R.T. Cullen, A. Searl, A.D. Jones and K. Donaldson, Inhalation of poorly soluble particles. II. Influence of particle surface area on inflammation and clearance, *Inhal. Toxicol.*, 2000, **12**(12), 1113–1126.

70. R.T. Cullen, C.L. Tran, D. Buchanan, J.M. Davis, A. Searl, A.D. Jones and K. Donaldson, Inhalation of poorly soluble particles, 1, *Differences in inflammatory response and clearance during exposure. Inhal. Toxicol.*, 2000, **12**(12), 1089–1111.

71. K.E. Driscoll, Role of inflammation in the development of rat lung tumors in response to chronic particle exposure, *Inhalation Toxicology*, 1996, **8**(suppl), 139–153.

72. G. Oberdorster, J. Ferin, S. Soderholm, R. Gelein, R. Baggs and P.E. Morrow, Increased pulmonary toxicity of inhaled ultrafine particles: due to overload alone, *Inhal. Toxicol.*, 1999, **38**, 295–302.

73. P.S. Gilmour, A. Ziesenis, E.R. Morrison, M.A. Vickers, E.M. Drost, I. Ford, E. Karg, C. Mossa, A. Schroeppel, G.A. Ferron, J. Heyder, M. Greaves, W. MacNee and K. Donaldson, Pulmonary and systemic effects of short-term inhalation exposure to ultrafine carbon black particles, *Toxicol. Appl. Pharmacol.*, 2004, **195**(1), 35–44.

74. L.C. Renwick, D. Brown, A. Clouter and K. Donaldson, Increased inflammation and altered macrophage chemotactic responses caused by two ultrafine particle types, *Occup. Environ. Med.*, 2004, **61**(5), 442–447.

75. D. Hohr, Y. Steinfartz, R.P. Schins, A.M. Knaapen, G. Martra, B. Fubini and P.J. Borm, The surface area rather than the surface coating determines the acute inflammatory response after instillation of fine and ultrafine TiO_2 in the rat, *Int. J. Hyg. Environ. Health*, 2002, **205**(3), 239–244.

76. R. Duffin, A. Clouter, D. Brown, C.L. Tran, W. MacNee, V. Stone and K. Donaldson, The importance of surface area and specific reactivity in the acute pulmonary inflammatory response to particles, *Ann. Occup. Hyg.*, 2002, **46**(Suppl 1), 242–245.

77. M.R. Wilson, J.H. Lightbody, K. Donaldson, J. Sales and V. Stone, Interactions between ultrafine particles and transition metals in vivo and in vitro, *Toxicol. Appl. Pharmacol.*, 2002, **184**(3), 172–179.

78. V. Stone, J. Shaw, D.M. Brown, W. MacNee, S.P. Faux and K. Donaldson, The role of oxidative stress in the prolonged inhibitory effect of ultrafine carbon black on epithelial cell function, *Toxicol. In Vitro*, 1998, **12**, 649–659.

79. I. Beck-Speier, N. Dayal, E. Karg, K.L. Maier, G. Schumann, H. Schulz, M. Semmler, S. Takenaka, K. Stettmaier, W. Bors, A. Ghio, J.M. Samet and J. Heyder, Oxidative stress and lipid mediators induced in alveolar macrophages by ultrafine particles, *Free Radic. Biol. Med.*, 2005, **38**(8), 1080–1092.

80. V. Stone, M. Tuinman, J.E. Vamvakopoulos, J. Shaw, D. Brown, S. Petterson, S.P. Faux, P. Borm, W. MacNee, F. Michaelangeli and K.

Donaldson, Increased calcium influx in a monocytic cell line on exposure to ultrafine carbon black, *Eur. Respir. J.*, 2000, **15**(2), 297–303.

81. D.M. Brown, K. Donaldson, P.J. Borm, R.P. Schins, M. Dehnhardt, P. Gilmour, L.A. Jimenez and V. Stone, Calcium and ROS-mediated activation of transcription factors and TNF-alpha cytokine gene expression in macrophages exposed to ultrafine particles, 5, *Am. J. Physiol. Lung. Cell. Mol. Physiol.*, 2004, **286**(2), L344–L353.

82. J. Tamaoki, K. Isono, K. Takeyama, E. Tagaya, J. Nakata and A. Nagai, Ultrafine carbon black particles stimulate proliferation of human airway epithelium via EGF receptor-mediated signaling pathway, 1, *Am. J. Physiol. Lung. Cell. Mol. Physiol.*, 2004, **287**(6), L1127–L1133.

83. C.R. Timblin, A. Shukla, I. Berlanger, K.A. Berube, A. Churg and B.T. Mossman, Ultrafine airborne particles cause increases in protooncogene expression and proliferation in alveolar epithelial cells, *Toxicol. Appl. Pharmacol.*, 2002, **179**(2), 98–104.

84. G. Oberdorster, Z. Sharp, A.P. Elder, R. Gelein, W. Kreyling and C. Cox, Translocation of inhaled ultratine particles to the brain, *Inhal. Toxicol.*, 2004, **16**, 437–445.

85. W.S. Tin-Tin, S. Yamamoto, S. Ahmed, M. Kakeyama, T. Kobayashi and H. Fujimaki, Brain cytokine and chemokine mRNA expression in mice induced by intranasal instillation with ultrafine carbon black, 32, *Toxicol. Lett.*, 2006, **163**(2), 153–160.

86. A. Radomski, P. Jurasz, D. onso-Escolano, M. Drews, M. Morandi, T. Malinski and M.W. Radomski, Nanoparticle-induced platelet aggregation and vascular thrombosis, 1, *Br. J. Pharmacol.*, 2005.

87. K. Donaldson and C.L. Tran, An introduction to the short-term toxicology of respirable industrial fibres, 5, *Mutat. Res.*, 2004, **553**(1–2), 5–9.

88. K. Nyberg, U. Johansson, I. Rundquist and P. Camner, Estimation of pH in individual alveolar macrophage phagolysosomes, *Exp. Lung Res.*, 1989, **15**(4), 499–510.

89. T.W. Hesterberg, W.C. Miiller, R.P. Musselman, O. Kamstrup, R.D. Hamilton and P. Thevenaz, Biopersistence of man-made vitreous fibers and crocidolite asbestos in the rat lung following inhalation, *Fundamental And Applied Toxicology*, 1996, **29**, 267–279.

90. J. Muller, F. Huaux, N. Moreau, P. Misson, J.F. Heilier, M. Delos, M. Arras, A. Fonseca, J.B. Nagy and D. Lison, Respiratory toxicity of multiwall carbon nanotubes, 1, *Toxicol. Appl. Pharmacol.*, 2005, **207**(3), 221–231.

91. A.D. Maynard, P.A. Baron, M. Foley, A.A. Shvedova, E.R. Kisin and V. Castranova, Exposure to carbon nanotube material: aerosol release during the handling of unrefined single-walled carbon nanotube material, *J. Toxicol. Environ. Health A*, 2004, **67**(1), 87–107.

92. C.W. Lam, J.T. James, R. McCluskey and R.L. Hunter, Pulmonary toxicity of single-wall carbon nanotubes in mice 7 and 90 days after intratracheal instillation, *Toxicol. Sci.*, 2004, **77**(1), 126–134.

93. D.B. Warheit, B.R. Laurence, K.L. Reed, D.H. Roach, G.A. Reynolds and T.R. Webb, Comparative Pulmonary Toxicity Assessment of Single-wall Carbon Nanotubes in Rats, *Toxicol. Sci.*, 2004, **77**(1), 117–125.

94. A.A. Shvedova, E.R. Kisin, R. Mercer, A.R. Murray, V.J. Johnson, A.I. Potapovich, Y.Y. Tyurina, O. Gorelik, S. Arepalli, D. Schwegler-Berry, A.F. Hubbs, J. Antonini, D.E. Evans, B.K. Ku, D. Ramsey, A. Maynard, V.E. Kagan, V. Castranova and P. Baron, Unusual inflammatory and fibrogenic pulmonary responses to single-walled carbon nanotubes in mice, 1, *Am. J. Physiol. Lung. Cell. Mol. Physiol.*, 2005, **289**(5), L698–L708.

95. N.A. Monteiro-Riviere, R.J. Nemanich, A.O. Inman, Y.Y. Wang and J.E. Riviere, Multi-walled carbon nanotube interactions with human epidermal keratinocytes, 8, *Toxicol. Lett.*, 2005, **155**(3), 377–384.

96. A.A. Shvedova, V. Castranova, E.R. Kisin, D. Schwegler-Berry, A.R. Murray, V.Z. Gandelsman, A. Maynard and P. Baron, Exposure to carbon nanotube material: assessment of nanotube cytotoxicity using human keratinocyte cells, *J. Toxicol. Environ. Health. A*, 2003, **66**(20), 1909–1926.

97. G. Jia, H. Wang, L. Yan, X. Wang, R. Pei, T. Yan, Y. Zhao and X. Guo, Cytotoxicity of carbon nanomaterials: single-wall nanotube, multi-wall nanotube, and fullerene, 4, *Environ. Sci. Technol.*, 2005, **39**(5), 1378–1383.

98. M. Bottini, S. Bruckner, K. Nika, N. Bottini, S. Bellucci, A. Magrini, A. Bergamaschi and T. Mustelin, Multi-walled carbon nanotubes induce T lymphocyte apoptosis, 1, *Toxicol. Lett.*, 2006, **160**(2), 121–126.

99. D. Cui, F. Tian, C.S. Ozkan, M. Wang and H. Gao, Effect of single wall carbon nanotubes on human HEK293 cells, *Toxicol. Lett.*, 2005, **155**(1), 73–85.

100. S.K. Manna, S. Sarkar, J. Barr, K. Wise, E.V. Barrera, O. Jejelowo, A.C. Rice-Ficht and G.T. Ramesh, Single-walled carbon nanotube induces oxidative stress and activates nuclear transcription factor-kappaB in human keratinocytes, 1, *Nano. Lett.*, 2005, **5**(9), 1676–1684.

101. J.C. Pache, Y.M. Janssen, E.S. Walsh, T.R. Quinlan, C.L. Zanella, R.B. Low, D.J. Taatjes and B.T. Mossman, Increased epidermal growth factor-receptor protein in a human mesothelial cell line in response to long asbestos fibers, *Am. J. Pathol.*, 1998, **152**(2), 333–340.

102. C.M. Sayes, F. Liang, J.L. Hudson, J. Mendez, W. Guo, J.M. Beach, V.C. Moore, C.D. Doyle, J.L. West, W.E. Billups, K.D. Ausman and V.L. Colvin, Functionalization density dependence of single-walled carbon nanotubes cytotoxicity in vitro, *Toxicol. Lett.*, 2006, **161**(2), 135–142.

103. C.M. Sayes, J.D. Fortner, W. Guo, D. Lyon and V.L. Colvin, The differential cytotoxicity of water-soluble fullerenes, *Nano. Letters.*, 2004, **4**(10), 1881–1887.

104. A. Isakovic, Z. Markovic, B. Todorovic-Markovic, N. Nikolic, S. Vranjes-Djuric, M. Mirkovic, M. Dramicanin, L. Harhaji, N. Raicevic, Z. Nikolic and V. Trajkovic, Distinct Cytotoxic Mechanisms of Pristine Versus Hydroxylated Fullerene, 1, *Toxicol. Sci.*, 2006.

105. S.M. Mirkov, A.N. Djordjevic, N.L. Andric, S.A. Andric, T.S. Kostic, G.M. Bogdanovic, M.B. Vojinovic-Miloradov and R.Z. Kovacevic,

Nitric oxide-scavenging activity of polyhydroxylated fullerenol, C_{60}(OH)24, 5, *Nitric Oxide*, 2004, **11**(2), 201–207.

106. A.A. Corona-Morales, A. Castell, A. Escobar, R. Drucker-Colin and L. Zhang, Fullerene C_{60} and ascorbic acid protect cultured chromaffin cells against levodopa toxicity, 13, *J. Neurosci. Res.*, 2003, **71**(1), 121–126.
107. N. Gharbi, M. Pressac, M. Hadchouel, H. Szwarc, S.R. Wilson and F. Moussa, [60]fullerene is a powerful antioxidant in vivo with no acute or subacute toxicity, 1, *Nano. Lett.*, 2005, **5**(12), 2578–2585.
108. Y.T. Lee, L.Y. Chiang, W.J. Chen and H.C. Hsu, Water-soluble Hexa-sulfobutyl[60]fullerene inhibit low-density lipoprotein oxidation in aqueous and lipophilic phases, 11, *Proc. Soc. Exp. Biol. Med.*, 2000, **224**(2), 69–75.
109. R. Hardman, A toxicologic review of quantum dots: toxicity depends on physicochemical and environmental factors, 2, *Environ. Health Perspect.*, 2006, **114**(2), 165–172.
110. J. Lovric, H.S. Bazzi, Y. Cuie, G.R. Fortin, F.M. Winnik and D. Maysinger, Differences in subcellular distribution and toxicity of green and red emitting CdTe quantum dots, *J. Mol. Med.*, 2005, **83**(5), 377–385.
111. B. Ballou, B.C. Lagerholm, L.A. Ernst, M.P. Bruchez and A.S. Waggoner, Noninvasive imaging of quantum dots in mice, 2, *Bioconjug. Chem.*, 2004, **15**(1), 79–86.
112. B. Dubertret, P. Skourides, D.j. Norris, V. Noireaux, A.H. Brivanlou and A. Libchaber, In vivo imaging of quantum dots encapsulated in phospholipid micelles, 1, *Science*, 2002, **298**(5599), 1759–1762.
113. D.R. Larson, W.R. Zipfel, R.M. Williams, S.W. Clark, M.P. Bruchez, F.W. Wise and W.W. Webb, Water-soluble quantum dots for multiphoton fluorescence imaging in vivo, 2, *Science*, 2003, **300**(5624), 1434–1436.
114. K.J. Soto, A. Carrasco, T.G. Powell, K.M. Garza and L.E. Murr, Comparative in vitro cytotoxicity assessment of some manufactured nanoparticulate materials characterised by transmission electron microscopy, *J. Nanoparticle Res.*, 2005, **7**, 145–169.
115. S.M. Hussain, K.L. Hess, J.M. Gearhart, K.T. Geiss and J.J. Schlager, In vitro toxicity of nanoparticles in BRL 3A rat liver cells, 1, *Toxicol. In vitro*, 2005, **19**(7), 975–983.
116. Z. Chen, H. Meng, G. Xing, C. Chen, Y. Zhao, G. Jia, T. Wang, H. Yuan, C. Ye, F. Zhao, Z. Chai, C. Zhu, X. Fang, B. Ma and L. Wan, Acute toxicological effects of copper nanoparticles in vivo, 1, *Toxicol. Lett.*, 2005.
117. C.A. Dick, D.M. Brown, K. Donaldson and V. Stone, The role of free radicals in the toxic and inflammatory effects of four different ultrafine particle types, *Inhal. Toxicol.*, 2003, **15**(1), 39–52.
118. K. Donaldson, P.H. Beswick and P.S. Gilmour, Free radial activity associated with the surface of particles: a unifying factor in determining biological activity, *Toxicol. Lett.*, 1996, **88**(1–3), 293–298.
119. C.A. Dick, D.M. Brown, K. Donaldson and V. Stone, The role of free radicals in the toxic and inflammatory effects of four different ultrafine particle types, *Inhal. Toxicol.*, 2003, **15**(1), 39–52.

Human Effects of Nanoparticle Exposure

LANG TRAN, ROB AITKEN, JON AYRES, KEN DONALDSON AND
FINTAN HURLEY

1 The Regulatory Issues

1.1 Nanosciences and Nanotechnologies per se

Since early 2003 there have been several publications, some at government level
across the world (*e.g.* the UK, Denmark, France, Canada, Netherlands),
highlighting issues relating to the imminent development of a range of nanotechnologies.[1-8] Issues involve two main areas: opportunities and human health
risks.[9-11] The first focuses on the new fields, uses and prospects for which
nanotechnology can be employed. The second focuses on concerns and challenges which so far have not been well investigated.[6]

Concerns have been expressed regarding the lack of knowledge about the
environmental and health implications of this fast-developing area[12-16] and
calls have been made for the development and implementations of stricter and
thorough regulations.[17,18]

In response to the Royal Society and Royal Academy of Engineering (2004)
report[3] on nanoscience and nanotechnology, the UK Government issued a
commitment to engage the public and the general community in early discussions
so that any aspirations and concerns can be addressed.[19] The report states that
precaution should be employed in the use of free engineered nanoparticles (NPs),
given the potential environmental and health concerns, and that steps should
be taken to identify, reduce or remove potential waste products containing
engineered NPs. In addition, nano-remediation using these materials should be
prevented until there is a fuller understanding of the risks.[19]

At the European Union (EU) level, a communication from the European
Commission, "Towards a European strategy for nanotechnology", was produced in May 2004[20] where a review of issues associated with the current
developments of nanosciences and nanotechnologies is presented. In this document, the Commission makes clear the importance of developments in this area

Issues in Environmental Science and Technology, No. 24
Nanotechnology: Consequences for Human Health and the Environment
© The Royal Society of Chemistry, 2007

and calls upon member states to develop proper risk assessments and generate adequate toxicological and ecotoxicological data and to evaluate potential human and environmental exposure.

This document was followed in June 2005 by a second communication, "Nanosciences and nanotechnologies (N&N): an action plan for Europe 2005–2009",[21] which highlighted that any environmental risks potentially associated with products and applications of N&N need to be addressed up front and throughout their lifecycle. There are therefore recommendations, at the EU level, for ensuring that the development of nanosciences and nanotechnology includes a proper and thorough evaluation of risks to health and the environment as an integral part of that development, especially in what concerns the use of manufactured free NPs.

1.2   Nanosciences and Nanotechnologies in Context of Dangerous Substances Generally

There is also a number of European legislations that have the objective of implementing laws regarding use of and exposure to dangerous substances generally. These include REACH[22] and the Council Directive 92/32/EEC, which amends Directive 67/548/EEC. It might appear that these are not relevant to NPs, because many materials currently manufactured at the nano-scale level are not new materials (*e.g.* carbon black). The situation is, however, more complex than that: because of their size and the ways in which they are used, NPs can possess specific physico-chemical properties that make them behave and interact differently from their parent materials when released into living systems. Therefore it is not possible to infer the safety of NPs by using information derived from the bulk parent material.[19]

Against this background, the Royal Society report[3] recommended that engineered NPs should be treated as new chemicals under UK and EU legislation so that appropriate testing procedures take place before their widespread use and environmental release. The UK Government agreed that these materials should undergo a thorough assessment before their general use.[19] Recommendations at international level also concur with these views. For example, the Royal Netherlands Academy of Sciences working group on Consequences of Nanotechnology proposed that more research should be done on the toxicological properties of nanoparticles and their kinetics in organisms and in the environment.[6]

2   Current Issues and Knowledge Gaps

The potential human health effects from exposure to NPs can be assessed properly only by combining the sources of information from toxicology, epidemiology and human challenge studies with available data on specific NP properties and their occupational or environmental exposure patterns. The state-of-the-art knowledge and issues regarding each of these important areas are addressed below.

2.1 Toxicology of Nanoparticles

Nanoparticles may be derived from three distinct sources: accidental release (generally combustion), manufactured/bulk and manufactured/engineered (see Table 1). The toxicological data available vary greatly between these sources, as shown in Table 1. One challenge for toxicologists is to determine whether the substantial knowledge derived from studies of the accidentally produced and bulk manufactured NPs can be generalised to engineered NPs.

Traditionally, with respect to the human hazard associated with inhalable or respirable particle exposure, the biological effects of a number of known particle types (*e.g.* silica, asbestos, particulate air pollution (PM_{10})) have been studied in animals and humans. Understandably, these studies have generally focused on particles entering the body via the lungs, rather than by dermal contact or ingestion. The adverse effects of engineered NPs on organs are currently better understood in the lungs than in other potential target organs (*e.g.* liver, cardiovascular system) because there is a paucity of toxico-kinetic data that would allow a proper evaluation of the NP dose that these organs receive.

The principal hypothetical routes of toxico-kinetics are shown in Figure 1. The potential of NPs to redistribute within cells, enter organelles (*e.g.* nucleus) or cross membranes and epithelia is of considerable interest. The small size of NPs makes it likely that they can cross membranes and even pass through nm-sized pores in membranes. There is evidence that some NPs have the ability to cross the epithelial barrier at the lung surface and enter the interstitium. Yet little is known of the ability of NPs to cross the intestinal epithelium or the skin. Also, there have been suggestions that NPs could gain access to the brain via the olfactory epithelium[23] and through the blood/brain barrier[24] for blood-borne NPs and this requires further investigation.

Biopersistence is the ability of particles or fibres to resist chemical dissolution. The concept of biopersistence has been used extensively in studying mineral fibres, as this is an important determinant of their pathogenicity, only long biopersistent fibres being pathogenic.[10] By analogy with asbestos fibres,

Table 1 Toxicological studies in different categories of NP.

Nanoparticles	Source	Potential exposure	Toxicology
Accidental	From combustion, road vehicles, fossil fuel cooking *e.g.* diesel	Low exposure everyone	Considerable data
Manufactured (1)	Bulk NPs in industry *e.g.* carbon black, TiO_2	High exposure workers	Some data
Manufactured (2)	Engineered NPs in the nanotechnology industry *e.g.* fullerenes, nanotubes, quantum dots	High exposure workers then low exposure everyone	Few data

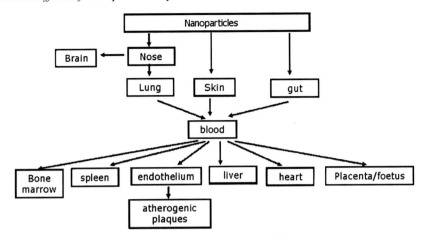

Figure 1 Hypothetical toxico-kinetic routes for NPs.

the biopersistence of nanotubes would likely play an important role in their pathogenicity. Currently, nothing is known of the ability of nanotubes to translocate to the pleura or peritoneal cavity following pulmonary deposition or whether they have any effects in these tissues.

If a specific NP were composed entirely of soluble material, its large surface-to-volume ratio means that it would rapidly undergo dissolution and would then no longer be considered a particle. There are no published data relating to the durability of any NPs.

While some particles of general respirable (*i.e.* not specifically nanoscale) size (*e.g.* TiO$_2$) are found to be relatively harmless at doses to which humans are exposed, other particle types (*e.g.* silica and asbestos) have been shown to induce inflammation leading to diseases such as fibrosis and cancer – see, for example, Donaldson *et al.*[25] and Donaldson and Tran.[26] In toxicology studies, the ability of such particles to induce toxic effects is related to a number of physico-chemical properties such as size, shape, chemical composition, surface reactivity, surface charge, solubility/biodurability.[27] For insoluble particles or an insoluble core of a complex particle, only the surface interacts with the biological system.[28] Therefore, the total surface area times its reactivity, in contact with the biological system, represents the "dose". For a given mass of any particular kind of particle, particle size determines the specific surface area, and so we would expect that particle size is important in toxicity and that smaller particle size means more harmful particles, per unit mass at the target organ. Indeed, some *in vitro* and *in vivo* studies of insoluble compounds or complex air-pollution particles have suggested that there may be key parameters, such as the specific surface area of the particles or their capacity to form free radicals, which control inflammation and toxicity (*e.g.*, Duffin *et al.*[28]). From thermodynamics this is logical, since the unit mass chemical reactivity (as well as thermodynamic instability) of a compound increases as particle size decreases, often resulting in changing polymorphs as a function of size.[29]

However, the possible pathogenic mechanisms induced by particle exposure are more complex and depend also on the route of exposure, host (response) susceptibility, and the specific physico-chemical properties of the individual particles (see *e.g.* Borm & Kreyling,[30]; review by Hoet *et al.*[31]). For example, the physico-chemical properties of the particle surface have been shown to play an important role in the biological effects in the lung and systemic circulation.[32] In these studies unmodified and negatively charged polystyrene NPs had no effect on blood thrombus formation whereas positively charged NPs enhanced thrombus formation when administered intravenously[33] or instilled intra-tracheally. Thus, by analogy with other particles which show differences in toxicity depending upon composition, we can anticipate that different types of NP will have different toxic potencies.

Additionally, there is likely to be a significant difference in the extent and mechanisms of toxicity of NPs dependent on their solubility under different biological and environmental conditions. For particles that have soluble materials (*e.g.* metals, organics) on the surface, there is potential for an extra toxic effect due to the soluble material. A lower solubility in water and biological fluids suggests that the biodurability will increase and this is highly relevant for exposure to quantum dots (which consist mainly of heavy metal NPs such as quantum cores of PbSe and CdSe and core-shell structure of CdSe/ZnS, HgSe/Fe and CdTe/CdS), gold NPs and fullerenes.

In contrast to these general NP properties, the actual stability of the particles within an organism depends on the local hydro-biochemical conditions, which vary greatly from organ to organ and by whether the NPs are located in an extra- or intracellular position. Studies have shown that the pH ranges from 1.5 to 8.4 in gastric fluid whereas the pH value in the lung and blood varies between 7 and 8 and is approximately 4 to 4.5 in compartments (phagosomes) of the macrophages (see *e.g.* review by Plumlee and Ziegler[34]). Similarly, redox conditions ($E_h = -\log [a_e]$) vary from highly oxidizing, *e.g.* at the skin or in the lung, to highly reducing in *e.g.* the intestine or interstitial positions respectively.

In summary the toxicology findings with nanoparticles point to some endogenous particle features that dictate toxicity, such as charge, biopersistence, total surface area, diffusible metals, ability to generate reactive oxygen species, *etc.* The importance of these features in determining toxicity is modified by host parameters encountered during deposition at portal of entry and translocation, *e.g.* pH, opsonising proteins. All of these factors combine to produce a total toxicity whose final effect depends on the target tissue.

2.2 NP Characterisation

It is well known that particles vary in their toxicity depending on size, shape, composition and surface properties. A comprehensive and accurate characterisation of NPs is therefore a prerequisite for any approach to investigate the potential hazard of specific NPs.

Several techniques have been applied for the physical and chemical characterisation of NPs because of the variety of structures (*e.g.* spheres, tubes, rods) and composition (*e.g.* organic, inorganic, double/modified layer, coating).[35] In addition to Infra Red (IR), Ultra Violet (UV), fluorescence (FL), Nuclear Magnetic Resonance (NMR) and X-ray spectroscopy,[36,37] techniques such as Scanning Electron Microscopy (SEM) and Transmission Electron Microscopy (TEM) are widely used for the size distribution characterisation of engineered NPs.[38–42] Other techniques, such as Gel Permeation Chromatography (GPC) and High Performance Liquid Chromatography (HPLC) coupled with UV and mass spectroscopy (MS), have also been applied in some cases.[41,43] Recently, the Flow Field Fractionation (FFF) technique has also been applied to the characterisation of NPs.[44,45]

Although there is an increasing usage of some NPs (*e.g.* carbon nanotubes and NPs for drug delivery applications) in medicine, only a few methods have been developed to identify NPs in biological tissues. These methods were based on techniques such as fluorescence,[46] TEM,[47] Inductively Coupled Plasma-Mass Spectroscopy (ICP-MS) and HPLC coupled with UV and MS.[48,49]

2.3 Epidemiology

2.3.1 Occupational Exposure.

Because engineering of NPs is relatively new, and long-term exposure may be needed before any adverse effects can be identified and attributed, there is little epidemiological evidence. On the other hand, there is extensive epidemiology on long-term and short-term exposure to dusts in many industries. The main problem is relevance, *i.e.* that these results do not necessarily generalise to the new nano-materials that are currently being developed precisely because they have new properties, and so health effects may also be different from those of existing, well-studied materials.

The most relevant occupational epidemiology studies are from the carbon black industry. Sorahan *et al.*[50] followed up an inconclusive early study by Hodgson and Jones[51] of mortality in male carbon black workers in five factories in Great Britain. Their main results, based on 1147 male manual workers employed for 12 months or more, showed a statistically significant excess of lung cancer compared to deaths expected based on death rates for England and Wales, with clear differences between factories. However, there was no evidence of a gradient with time since first employment and, after adjustment for factory and other risk factors, no clear evidence of a trend with cumulative exposures, *i.e.* the study did not show any clear evidence that the excess was work related. However, data limitations imply that the possibility of some effect cannot be ruled out. Mortality overall from non-malignant respiratory disease and circulatory disease was similar to that expected, based on age-specific death rates for men in England and Wales.

A series of papers on the respiratory health of workers in the European carbon black industry, including chest radiographs[52,53] and lung function and respiratory symptoms,[52,54] show some adverse effects of exposure to carbon black dust on respiratory health. Results from two cross-sectional surveys (1991–92 and 1994–95) of lung function and respiratory symptoms in overlapping though non-identical groups of about 2000 workers at 19 and 16 plants, respectively, showed some consistent evidence of effects on respiratory symptoms and lung function (FEV_1 and mid-flow rather than FVC), which appear to be associated principally with recent rather than cumulative exposure.[54] There was, in addition, clear radiological evidence of dust retention, and relationships of radiological endpoints with cumulative dust exposure, though little evidence of what is usually called radiologically identified disease (*i.e.* Category 2 or more small opacities). Qualitatively similar results were found in studies cross-sectionally[52] and longitudinally.[53]

The main implications of these results are reassuring with regard to carbon black. Small methodological variations would surely give rise to somewhat different results, but not to markedly different conclusions, *i.e.* it is highly unlikely that severe work-related risks remain unidentified. Unfortunately, there is limited information about the size of the primary particles in the carbon black studies, and the toxicity of NPs is affected by other factors also. Thus, it is not possible to generalise from these results.

Epidemiology following exposure to other particles does not help greatly. Mortality and morbidity results from studies of workers producing titanium dioxide (TiO_2) were reviewed by Hext et al.[55] Overall, these results are similar in nature to those for carbon black. However, nanoparticle TiO_2 is a very small fraction of the overall market and the results applied particularly to coarser TiO_2. Welders are exposed to very fine particles. There is a huge literature on mortality and morbidity of welders. However, the effects depend on exactly what metals are being welded and so it is difficult to extrapolate to the new nanotechnology industries. We note, however, that some of the effects may be immediate (acute). There is an extensive literature confirming excess cancer mortality (lung cancer, mesothelioma) and respiratory morbidity following occupational exposure to asbestos, as well as studies of other fibres. This may be relevant, because of the fibrous shape of some new nanomaterials. Characteristics determining toxicity are length greater than 10–15 μm, diameter less than 3 μm, and solubility in the lung milieu, with a further contribution from surface properties. Nanotubes with these characteristics might be expected to show similar hazards; risks would depend on dose inhaled and retained. Finally, most studies of particles in outdoor air are based on mass concentrations of particulate matter (PM). There are some studies of particle number, almost all from Germany or from Finland. These are more relevant to nanoparticles, e.g. from traffic. They show adverse effects associated with the ultrafine fraction of respirable particles on mortality and on individuals with asthma;[56] the effect of ultrafine particles seems to be delayed by a few days.

2.3.2 Environmental Exposure. It is well known that daily variations in particles in outdoor air are associated with a very wide range of adverse health effects, including increased cardio-respiratory mortality, increased hospital admissions from respiratory and cardiac causes, increases in visits to primary care physicians (GPs), increases in work and school absenteeism, exacerbations of asthma and changes in detailed measures of respiratory and cardiac health such as lung function and heart rate variability. Furthermore, long-term exposure to outdoor particles has been shown to be associated with earlier mortality (reduced life expectancy) in adults and with increased mortality in infants.

These results are based on studies of mass concentrations of ambient particles, for example measured as PM_{10} or $PM_{2.5}$, principally because mass concentrations in PM_{10}, and increasingly in $PM_{2.5}$ also, are measured routinely in a wide range of locations in North America, in Europe (including the UK) and elsewhere. It has long been conjectured however that small particles from combustion sources have a key role in driving the PM-health relationships. This has led to an interest in epidemiology which can differentiate the effects of ultrafine from larger respirable particles, using, for example, measure of particle size.

There is as yet a limited number of relevant epidemiological studies, almost all from Germany or from Finland. One of the German studies examines mortality.[57] The other studies examine panels of people with impaired health, mostly pollution-related changes in the respiratory health of people with asthma, though cardiovascular health endpoints have been studied also.

Papers up to and including those published in 2002 have been reviewed in detail by Morawska *et al.*[56] They conclude the following:

- There are adverse effects associated with the ultrafine fraction of respirable particles.
- Those effects are shown on mortality in the general population, and on panels of individuals with asthma or other impaired health.
- The studies suggest a more-or-less immediate effect of fine particles as conventionally measured ($PM_{2.5}$), whereas the effect of ultrafine particles seems to be delayed by a few days.
- This suggests that fine particles are insufficient as a surrogate for ultrafine particles.

In addition, there have been more recent publications which add information to the role of nanoparticles using the number metric as a surrogate measure. A study from Rome of out-of-hospital sudden death[58] showed that death was related to particle numbers in the nano-size range, particularly on the day of death, but to a greater degree than either particle mass or carbon monoxide concentrations. On the other hand, a study of elderly individuals with coronary artery disease suggested that particle mass ($PM_{2.5}$) was the more important driver of health effects although particle numbers did relate to activity restriction.[59]

These studies support the toxicological findings, earlier, that the effects of dust depend not only on its size, but also on what kind of dust it is. In order to draw general conclusions it is important to characterise as well as possible the nature and size distribution of the aerosols to which workers are exposed, and to track in sufficient detail workers' time spent in these conditions. In particular, epidemiology has limited value for assessing the effects of NPs unless it can differentiate (in time and space) between exposure to nanoparticles and exposure to particles from other sources. Finally, as ever, study power is a factor – numbers of subjects, length of follow-up, variation in exposure – implying a need for multi-centre studies with associated issues of design and management.

2.4 Human Challenge Studies

Human exposure studies assess biological responses to controlled levels of specific air mixtures by a range of approaches, allowing precise control of delivered concentrations, adjustment of ventilation rates and measurement of a range of biological responses. These studies are ideal for assessing acute (*i.e.* immediate to 24 hours) responses to pollutants but not for assessing the effects of long-term exposure to specific substances.

The range of biological responses studied will vary from invasive procedures such as biopsy of the lung by bronchoscopy to less invasive assessment of systemic responses from blood samples. Timing of response in relation to exposure is also crucial but a key issue is determining what could be regarded as a "normal", physiological response rather than a patho-physiological (abnormal) response. Understanding these differences is crucial when considering exposures that may have subtle yet important effects after short-term exposures.

As there are no studies specifically of manufactured NPs, the available information comes from studies of particles derived from the internal combustion engine or from laboratory-generated particles, whose content is part of that seen in ambient particles. The great majority of published human exposure studies considered a source of particles for which either the exact particle size range was unknown or where the range included, but was not limited to, the nanoparticle range (*e.g.* exposures to diesel exhaust particles (DEPs) and Concentrated Ambient Particles (CAPs)). The only studies of particles solely in the nanoparticle range are those where the particles are specifically generated in the laboratory (usually using an electrical spark generator) and thus usually comprise a single type of particle (*e.g.* carbon, iron or zinc).

For a range of reasons, including availability of volunteers and ethical concerns about exposing individuals with moderate to severe disease to a potentially risky substance, most subjects used in these studies have been younger, healthy volunteers, although some studies have exposed subjects with mild asthma. Only two studies have exposed individuals with significant cardiac disease[60] or chronic lung disease.[61]

There have been studies of DEP exposure giving adequate information on particle size distribution.[62–66] Most delivered moderately high concentrations in mass terms (100–300 µg m^{-3}). These studies broadly suggested that DEPs can cause a neutrophilic inflammatory response in both normal and asthmatic subjects, probably mediated through IL-8, with some evidence of endothelial activation especially in the asthmatic subjects. There are two reports of CAP exposures, which suggest an airway neutrophilic response perhaps with an endothelial response.[67]

Exposure to zinc oxide fume produces conflicting results. Spark-generated zinc oxide produces no effects on any inflammatory marker.[68] Zinc oxide fume at high exposures (up to 37 $\mu g\ m^{-3}$) produced systemic effects typical of metal fume fever, increased plasma IL-6 and a dose-related increase in BAL neutrophils.[69,70]

Exposure to sulfuric acid particles (in the nanometre range) has shown no effects on lung function or inflammatory markers.[71,72] However, one study of the effects of ultrafine sulfuric acid (1000 $\mu g\ m^{-3}$) on response to allergen exposure[73] showed enhancement of the allergen response in the airways. Two studies of ultrafine carbon exposure showed an immediate effect on cardiac autonomic control in patients with cardiac disease[60] and a suggestion that on exercise there was shortening of the Q-T interval of the ECG in older subjects.[74]

Most reported human challenge work relates to particles in ambient air. These particles constitute an insoluble, solid core (in the case of ambient particles, carbon) or comprise soluble substances (*e.g.* sulfuric acid or sulfates). The solid particles form a base for the carriage of other molecules of greater or lesser bio-activity so an approach to decide on mechanisms of such complex structures needs either to take the pragmatic view that the whole particle should be assessed or to study the individual components in isolation. In addition, the dose and the physico-chemical characteristics of the particles are fundamental drivers of biological responses. Now that particles in the ultrafine range are recognised as important, mass may be an inappropriate index of dose as NPs are very light but have a vast surface area. There is a need to be able to develop systems for delivering nanomaterials in these studies in sufficient dose but using surface area or number as the dosing metric.

3 Discussion: Risk Assessment of Engineered NPs

The need for a proper risk assessment of NPs is well accepted by the European scientific community.[75] The application of risk assessment to NPs is a complex issue and only a few initial attempts have been reported so far.[76] This is mainly due to methodological and knowledge gaps that are still present: quantitative risk assessment implies that sources of NPs as well as human exposure and associated health effects are well characterised, and this implies a need for evidence beyond what is available at present.

A tentative risk assessment can however be structured into four steps: source identification, exposure assessment, hazard assessment and risk characterisation.[76] We consider these steps in turn.

For source identification, there is currently little information on the potential release of NPs during the lifecycles of different applications.[76] The probability and the rate of release are strictly dependent on techniques of production and types of application, which are still heterogeneous and changing rapidly; *e.g.*, it has recently been found that production and storage in liquid can reduce the airborne exposure.[3]

For exposure assessment, current knowledge of NP behaviour in different environmental media is still uncertain. The behaviour will be primarily related to the physico-chemical form of the NPs: dispersed, adsorbed to other substrate, aggregated or taken up by cells.[77] For example, C_{60} (fullerene) is a hydrophobic nanomaterial, but it can form an aqueous suspended stable colloidal species in water.[78] On the other hand, in water elementary carbon particles or nanotubes show a tendency to aggregate.[79] The mobility of NPs in aqueous environments is a function of particle transport, transformation and removal mechanism.[80] Fluid flow, gravity and diffusion are the primary mechanisms for transport. Ionic strength and pH may be relevant parameters that affect the NP mobility in water. Initial studies of environmental nanotechnologies indicate that iron NPs, used for cleaning up contaminated soil and groundwater, can travel with groundwater over a distance of 20 metres and remain reactive for 4–8 weeks.[48] Nanoparticle surface properties (such as large surface area per unit mass, crystalline structure, anionic surface functional groups) make them ideally suited to carry or sequester adsorbed hazardous chemicals species (*e.g.* heavy metals), thus enhancing the bioavailability of pollutants.

For hazard assessment, evidence of NP toxicity only begins to emerge from the toxicology, epidemiology and human challenge studies mentioned above. Although the findings from epidemiology and human exposure studies have not provided evidence of serious risks to health from the few possibly relevant NPs studied, we think it would be wrong to conclude that engineered NPs are safe. Rather, the evidence from toxicology studies strongly suggests that there may be wide variations in the toxicity of engineered NPs, according to their physico-chemical properties. This includes the possibility that some engineered NPs may represent a serious hazard to health.

It follows that identification of which NPs are in fact hazardous is a priority. This implies a comprehensive international programme of toxicological research, to clarify further the role of various physico-chemical characteristics, individually and in combination, in determining toxicity; and a parallel programme of toxicity testing of specific materials. It is moreover important that, since NPs by definition have different properties from the same chemical in larger scale, the exposure of those working with such new materials must be characterised in a way that reflects their exposure to NPs specifically and not simply to particles in general. This applies also to human exposure studies. Finally, cohorts of workers exposed to industrial NPs and the characterisation of NPs to which they are exposed may provide useful indicators of the amount of exposure at which adverse effects might expect to be detected. In any major industry in which exposure to new NPs is likely, consideration should be given, in the UK perhaps under the COSHH processes, to establishing a cohort of such workers.

Consumer exposure to NPs is likely to increase in the future although there is still a knowledge gap on the averse effects from dermal and ingestion exposure to NPs. Future *in vitro* toxicology tests will help to elucidate the mechanisms by which NPs exert their toxicity on cells while human challenge studies will continue to inform research on the human dosimetry and short-term effects. Current and future studies in these fields will undoubtedly narrow the knowledge gaps regarding consumer exposure also.

For risk characterisation, the current level of uncertainty in exposure and hazard, and the need to integrate across evidence from different kinds of studies, suggests the use of a Weight-of-Evidence (WoE) approach. This approach has received a large measure of consensus as a methodology in recent ecological risk assessment studies.[81–83] The WoE approach does not estimate the risk simply on the basis of predicted exposure concentrations and dose-response curves, but integrates exposure and response lines of evidence in order to assure a better understanding of the dynamics of a complex system. Its proponents claim that a WoE approach can overcome some of the limitations of the traditional approaches because it has the potential substantially to reduce uncertainty associated with risk assessment and improve management decisions.[81] We think that a WoE approach should be considered as a way forward for a rational assessment of risk posed by exposure to engineered nanoparticles.

References

1. S. Bergeron and E. Archambault, Canadian stewardship practices for environmental nanotechnology. Prepared for Environment Canada, 2005.
2. Small sizes that matter: Opportunities and risks of Nanotechnologies. Munich: Allianz, 2005.
3. The Royal Society and Royal Academy of Engineering, Nanoscience and nanotechnologies: opportunities and uncertainties, RS Policy document 19/04, UK, 2004.
4. A.H. Arnall, Future technologies, today's choices. Nanotechnology, artificial intelligence and robotics; a technical, political and institutional map of emerging technologies. Report for the Greenpeace Environmental Trust, UK, 2003.
5. R. Corriu, P. Nozières and C. Weishbuch, Nanosciences – Nanotechnologies. Rapport Science et Technologie, No. 18, realisé avec l'Academie des Technologies. France, 2004.
6. R. Van Este and I. Van Keulen, in *Industrial Application of Nanomaterials – Chance and Risks: Technological Analysis*, ed. W. Luther, VDI, Dusseldorf, 2004.
7. F. Durrenberger and K. Hohener, Overview of completed and ongoing activities in the field: safety and risks of nanotechnology, Technologie Management, TEMAS AG, Switzerland, 2004.
8. Willems and Van den Wildenberg, Nanoroadmap project. State of the art overview and forecasts based on existing information of nanotechnology in the field of NP. Co-funded by the 6th Framework Programme of the European Commission, 2004.
9. A. Helland, MSc thesis, Lund University, Sweden, 2004.
10. K. Donaldson, V. Stone, C.L. Tran, W. Kreyling and P.J.A. Borm, *Occup. Environ. Med.*, 2004, **61**, 727–728.
11. E. Hood, *Environ. Health Perspect.*, 2004, **112**, A741–A749.
12. G.H. Reynolds, *Environ. Law Reporter*, 2001, **31**, 10681–10688.

13. M. Kosal, *B. Atom Sci.*, 2004, **60**(5), 38–47.
14. ETC Group, No small matter! Nanotech particles penetrate living cells and accumulate in animal organs, ETC Communiqué, May/June, Issue #76, 2002.
15. ETC Group, The Big Down: from genomes to atoms. Atomtech: technologies converging at the nanoscale, ETC Group, Canada, 2003.
16. ETC Group, Nano's troubled waters: latest toxic warning shows nanoparticles cause brain damage in aquatic species and highlights need for a moratorium on the release of new NP, ETC Group, Canada, 2004.
17. L. Sheremata and A.S. Daar, *Health Law Rev.*, 2005, **12**, 74–77.
18. N. Jacobstein and G.H. Reynolds, *Foresight guidelines version 4.0: self-assessment scorecards for safer development of nanotechnology*, Foresight Institute, USA, 2004.
19. HM Government, Response to the Royal Society and the Royal Academy of Engineering Report: 'Nanoscience and nanotechnologies: opportunities and uncertainties', 2005.
20. European Commission, Towards a European Strategy for Nanotechnology, 2004. Report available on http://cordis.europa.eu/nanotechnology/actionplan.htm.
21. European Commission, Nanosciences and Nanotechnologies (N&N): an action plan for Europe 2005–2009, 2005. Report available on http://ec.europa.eu/research/industrial_technologies/lists/documentlibrary_en.html.
22. REACH legislation http://eurlex.europa.eu/LexUriServ/site/en/oj/2006/1_396/1_39620061230en08500856.pdf.
23. G. Oberdörster, Z. Sharp, V. Atudorei, A. Elder, R. Gelein, W. Kreyling and C. Cox, *Inhal. Toxicol.*, 2004, **16**, 437–445.
24. J. Kreuter, *J. Nanosci. Nanotechno.*, 2004, **4**, 484–488.
25. K. Donaldson, D. Brown, A. Clouter, R. Duffin, W. MacNee, L. Renwick, C.L. Tran and V. Stone, *J. Aerosol. Med.*, 2002, **15**, 213–220.
26. K. Donaldson and C.L. Tran, *Inhal.Toxicol.*, 2002, **14**, 5–27.
27. W.G. Kreyling, M. Semmler and W. Moller, *J. Aerosol Med.*, 2004, **17**, 140–152.
28. R. Duffin, C.L. Tran, A. Clouter, D.M. Brown, W. MacNee, V. Stone and K. Donaldson, *Ann. Occup. Hyg.*, 2002, **46**(Suppl. 1), 242–245.
29. A. Navrotsky, in *Nanoparticles and the Environment*, ed. J.F. Banfield, A. Navrotsky and P.H. Ribbe, The Mineralogical Society of America, Reviews in Mineralogy and Geochemistry, 2001, vol. 44, pp. 73–103.
30. P.J. Borm and W. Kreyling, *J. Nanosci. Nanotechnol.*, 2004, **4**, 521–531.
31. P.H.M. Hoet, I. Bruske-Hohfeld and O.V. Salata, *J. Nanobiotechnol.*, 2004, **2**, 12.
32. J. Hamoir, A. Nemmar, D. Halloy, D. Wirth, G. Vincke, A. Vanderplasschen, B. Nemery and P. Gustin, *Toxicol. Appl. Pharmacol.*, 2003, **190**, 278–285.
33. A. Nemmar, P.H. Hoet, B. Vanquickenborne, D. Dinsdale, M. Thomeer, M.F. Hoylaerts, H. Vanbilben, L. Hortelmans and B. Nemery, *Circulation*, 2002, **105**, 411–414.

34. G.S. Plumlee and T.L. Ziegler, in *Environmental Geochemistry*, ed. B.S. Loller, Treatise on Geochemistry, 9, 2004, The United States Geological Survey, 263–310.
35. T. Satoshi, F. Minoru and H. Shinji, *Phys. Rev. B*, 2002, **66**(24), 245424.
36. S.I. Seok and J.H. Kim, *Mater. Chem. Phys.*, 2004, **86**, 176–179.
37. J. Fan, J. Lu, R. Xu, R. Jiang and Y. Gao, *J. Colloid Interface Sci.*, 2003, **266**, 215–218.
38. M. Niederberger, G. Garnweiter, F. Krumeirk, R. Nerper, H. Colfen and M. Antonietti, *Chem. Mater.*, 2004, **16**, 1202–1208.
39. S. Utsunomiya, K.A. Jensen, G.J. Keeler and R.C. Ewing, *Environ. Sci. Technol.*, 2002, **36**, 4943–4947.
40. S. Utsunomiya, K.A. Jensen, G.J. Keeler and R. C Ewing, *Environ. Sci. Technol.*, 2004, **38**, 2289–2297.
41. D. Farrell, S.A. Majetich and J.P. Wilcoxon, *J. Phys. Chem.*, 2003, **107**, 11022–11030.
42. A. Goel, P. Hebgen, J.B. Vander Sande and J.B. Howard JB, *Carbon*, 2002, **40**, 177–182.
43. W.P. Peng, Y. Cai, Y.T. Lee and H.-C. Chang, *Int. J. Mass Spectrom.*, 2003, **229**, 67–76.
44. M. Montalti, L. Prodi, N. Zaccheroni, A. Zattoni, P. Reschiglian and G. Falini, *Langmuir*, 2004, **20**, 2989–2991.
45. N. Tagmatarchis, M. Prato and D.M. Guldi, *Physica E*, 2005, **29**(3), 546–550.
46. C.S. Yang, C.H. Chang, P.J. Tsai, W.Y. Chen, F.G. Tsen and L.W. Lo, *Anal. Chem.*, 2004, **76**, 465–471.
47. W. Hartig, B.R. Paulke, C. Varga, J. Seeger, T. Harkany and J. Kazca, *Neurosci. Lett.*, 2003, **338**, 174–176.
48. W.X. Zhang, *J. Nanopart. Res.*, 2003, **5**, 323–332.
49. F. Moussa, M. Pressac, E. Genin, S. Roux, F. Trivin, A. Rassat, R. Ceolin and H. Szwarc, *J. Chrom. B.*, 1997, **696**, 153–159.
50. T. Sorahan, L. Hamilton, M. van Tongeren, K. Gardiner and J.M. Harrington, *Am. J. Ind. Med.*, 2001, **39**, 158–170.
51. J.T. Hodgson and R.D. Jones, *Arch. Environ. Health*, 1985, **40**, 261–268.
52. K. Gardiner, N.W. Trethowan, J.M. Harrington, C.E. Rossiter and I.A. Calvert, *Brit. J. Ind. Med.*, 1993, **50**, 1082–1096.
53. M.J. van Tongeren, K. Gardiner, C.E. Rossiter, J. Beach, P. Harber and M.J. Harrington, *Eur. Respir. J.*, 2002, **20**, 417–425.
54. K. Gardiner, M. van Tongeren and M. Harrington, *Occup. Environ. Med.*, 2001, **58**, 496–503.
55. P.M. Hext, J.A. Tomenson and P. Thompson, *Ann. Occup. Hyg.*, 2005, **49**, 461–472.
56. L. Morawska, M.R. Moore and Z.D. Ristovski, Health impacts of ultra-fine particles – desktop literature review and analysis, Section 5.4, 2004. Available at http://www.deh.gov.au/atmosphere/airquality/publications/health-impacts/priorities.html.

57. H.E. Wichmann, C. Spix, T. Tuch, G. Wolke, A. Peters, J. Heinrich, W.G. Kreyling and J. Heyder, *Health Effects Institute*, 2000, **98**, 5–86; discussion 87–94.
58. F. Forastiere, M. Stafoggia, S. Picciotto, T.D. Bellander D'Ippoliti, T. Lanki, S. von Klot, F. Nyberg, P. Paatero, A. Peters, J. Pekkanen, J. Sunyer and C. Perucci, *Am. J. Respir. Crit. Care Med.*, 2005, doi:10.1164/rccm.200412-1726OC.
59. J.J. de Hartog, G. Hoek, A. Peters, K.L. Timonen, A. Ibald-Mulli, A.B. Brunekreef, J.P. Heinrich Tiittanen, J.H. van Wijnen, W. Kreyling, M. Kulmala and J. Pekkanen, *Am. J. Epidemiol.*, 2003, **157**, 613–623.
60. H.C. Routledge, S. Manney, R.M. Harrison, J. Ayres and J.N. Townend, *Heart*, 2005, May 27, Epub. ahead of print.
61. H. Gong, W.S. Linn, K.W. Clark, K.R. Anderson and M.D. Geller, Sioutas. Respiratory responses to exposures with fine particulates and nitrogen dioxide in the elderly with and without COPD, *Inhalation Toxicology*, 2005, **17**, 123–132.
62. B. Ruddell, M.-C. Ledin, U. Hammarstrom, N. Stjernberg, B. Lundback and T. Sandstrom, *Occup. Environ. Med.*, 1996, **53**, 658–662.
63. S. Salvi, A. Blomberg, B. Rudell, F. Kelly, T. Sandrom, S. T. Holgate and A. Frew, *Am. J. Respir. Crit. Care Med.*, 1999, **159**, 702–709.
64. C. Nordenhall, J. Pourazar, A. Blomberg, J.O. Levin, T. Sandstrom and E. Adelroth, *Eur. Respir. J.*, 2000, **15**, 1046–1051.
65. S.S. Salvi, C. Nordenhall, A. Blomberg, B. Rudell, J. Pourazar, F.J. Kelly, S. Wilson, T. Sandstrom, S.T. Holgate and A.J. Frew, *Am. J. Respir. Crit. Care Med.*, 2000, **161**, 550–557.
66. J.A. Nightingale, R. Maggs, P. Cullinan, L.E. Donelly, D.F. Rogers, R. Kinnersley, K. Fan Chung, P.J. Barnes, M. Ashmore and A. Newman-Taylor, *Am. J. Respir. Crit. Care Med.*, 2000, **162**(1), 161–166.
67. H. Gong Jr., C. Sioutas and W.S. Linn, *Effects of Health Institute*, 2003; **118**, 1–36; discussion 37–47.
68. W.S. Beckett, D.F. Chalupa, A. Pauly-Brown, D.M. Speers, J.C. Stewart, M.W. Frampton, M.J. Utell, L.S. Huang, C. Cox, W. Zareba and G. Oberdörster, *Am. J. Respir. Crit. Care Med.*, 2005, **171**(10), 1129–1135.
69. W.G. Kuschner, A. D'Alessandro, S.F. Wintermeyer, G.H. Wong, H.A. Boushey and P.D. Blank, *J. Invest. Med.*, 1995, **43**, 371–378.
70. W.G. Kuschner, H. Wong and A. D'Alessandro, *Environ. Health Persp.*, 1997, **105**, 1234–1237.
71. M.W. Frampton, K.Z. Voter, P.E. Morrow, N.J. Robert Jr., D.J. Culp, C. Cox and M.J. Utell, *Am. Rev. Respir. Dis.*, 1992, **146**, 626–632.
72. W.S. Tunnicliffe, D. Mark, J.G. Ayres and R.M. Harrison, *Eur. Resp. J.*, 2001, **18**, 640–647.
73. W.S. Tunnicliffe, F.J. Kelly, C. Dempster, R.M. Harrison and J.G. Ayres JG, *Occup. Environ. Med.*, 2003, **60**, e15.
74. M.W. Frampton, M.J. Utell, W. Zareba, G. Oberdörster, C. Cox, L.S. Huang, P.E. Morrow, F.E., Lee, D. Chalupa, L.M. Frasier, D.M., Speers,

and J. Stewart, *Res. Rep. Health Eff. Inst.*, 2004, **126**, 1–47; discussion 49–63.
75. Better Regulation Taskforce http://www.brc.gov.uk/downloads/pdf/ scienceresponse.pdf.
76. European Commission, Nanotechnologies: a preliminary risk analysis on the basis of a workshop organized in Brussels on 1–2 March 2004 by the Health and Consumer Protection Directorate General of the European Commission, 2004. Report available on http://europa.eu.int/comm/health/ ph_risk/events_risk_en.htm.
77. Swiss Re, *Carbon*, 2004, **43**, 1984–1989.
78. G.V. Andrievsky, N.O. Mchedlov-Petrossyan and V.K. Klochkov, *J. Chem. Soc. Faraday Trans.*, 1997, **93**, 4343–4346.
79. V. Colvin, *Nature Biotech.*, 2003, **21**, 1166–1170.
80. M.R. Wiesner, H. Lecoanet and M. Cortalezzi, Nanomaterials, Sustainability and Risk Minimization – Introduction to IWA International Conference on Nano and Microparticles in Water and Wastewater Treatment. Zurich, Switzerland, 22–24 September 2003.
81. P.M. Chapman, B.G. McDonald and G.S. Lawrence, *Human Ecol. Risk Assess.*, 2002, **8**, 1489–1516.
82. G.A. Burton Jr, P.M. Chapman and E.P. Smith, *Human Ecol. Risk Assess.*, 2002, **8**, 1657–1673.
83. C. Menzie, M.H. Henning, J. Cura, K. Finkelstein, J.H. Gentile, J. Maughan, D. Mitchell, S. Petron, B. Potocki, S. Svirsky and P. Tyler, *Human Ecol. Risk Assess.*, 1996, **2**, 277–304.

Nanoparticle Safety – A Perspective from the United States

ANDREW D. MAYNARD

1 Introduction

Nanotechnology has variously been described as a transformative technology, an enabling technology and the next technological revolution. Even accounting for a certain level of hype, a heady combination of high-level investment, rapid scientific progress and exponentially increasing commercialisation point towards nanotechnology having a significant impact on society over the coming decades. However, enthusiasm over the rate of progress being made is being tempered increasingly by concerns over possible downsides of the technology, including unforeseen or poorly managed risk to human health.[1-4] Real and perceived adverse consequences in areas such as asbestos, nuclear power and genetically modified organisms have engendered increasing scepticism over the ability of scientists, industry and governments to ensure the safety of new technologies. As nanotechnology moves towards widespread commercialisation, not only is the debate over preventing adverse consequences occurring at an unusually early stage in the development cycle, but it is expanding beyond traditional knowledge-based risk management to incorporate public perception, trust and acceptance.[5-8]

Within this context, the long-term success of "nanotechnologies" (referring to the many specific applications and implementations of nanotechnology) will depend on rational, informed and transparent dialogue aimed at understanding and minimising the potential adverse implications to human health and the environment. A central question within this dialogue, and one that has been raised in the popular media as well as in the peer-reviewed press, is "how safe is nanotechnology?".[2,3,9-13] Of course, "safe" is relative, and needs to be understood in the context of specific nanotechnologies and their applications. Nevertheless, there is an increasing desire to understand the potential risks associated with emerging technologies, and how they might be managed to ensure the benefits outweigh any downsides. This paper provides a perspective on current

Issues in Environmental Science and Technology, No. 24
Nanotechnology: Consequences for Human Health and the Environment
© The Royal Society of Chemistry, 2007

activities within the USA addressing "safe" nanotechnology – including potential risks associated with the production and use of engineered nanoparticles.

2   The US National Nanotechnology Initiative

In many ways, the USA has been a leader in international interest in nanotechnology. Through the late 1990s an Interagency Working Group on Nanotechnology (IWGN) was active in promoting research and development in this area.[14,15] In 2001, the Clinton administration raised nanoscience and technology to the level of a national initiative, and the US National Nanotechnology Initiative (NNI) was formally established. Under the auspices of the National Science and Technology Council (NSTC), the NNI was to coordinate nanotechnology-related research across federal agencies. Although the NNI lacked (and still lacks) authority to allocate research funding, its formation stimulated research across the federal government, and has led to an increase in nano-specific research and development funding, from $464 million in 2001 to an estimated $1301 million in 2006.[16]

The role of the NNI was further formalised in 2003 through the signing of the 21st Century Nanotechnology Research and Development Act.[17] Through the Nanoscale Science, Engineering and Technology (NSET) subcommittee of NSTC, specific charges were given to agencies operating within the NNI. As well as addressing new research underpinning the application and commercialisation of nanotechnology, the Act also raised the issue of societal impact. Section 2(B) 10 of the Act covers responsibility for

"ensuring that ethical, legal, environmental, and other appropriate societal concerns, including the potential use of nanotechnology in enhancing human intelligence and in developing artificial intelligence which exceeds human capacity, are considered during the development of nanotechnology".

This is to be achieved through establishing appropriate research programs and interdisciplinary research centres, integrating research on societal, ethical and environmental concerns with nanotechnology research and development, and providing for public input and outreach.[17] Although there is a heavy emphasis on ethical, legal and other social implications, the Act does encompass environmental, safety and health implications of emerging nanotechnologies.

A commitment to addressing potential societal impacts of nanotechnology within the US federal government was further emphasised with the publication of the NNI strategic research plan in 2004.[18] The plan outlined four broad goals, and seven Program Component Areas, forming the basis of a strategic and reviewable plan. While fundamentally focused on the development and applications of nanotechnology, a commitment to address the

"societal dimensions related to the development of new technologies, including the potential implications for health and the environment, and the importance of dialogue with the public" was emphasised by John H. Marburger III, the Director of the Office of Science and Technology Policy.[18] In the plan, this commitment was supported by one specific goal addressing the responsible development of nanotechnology, and a broader Program Component Area addressing societal dimensions of nano-technology.

Since 2003, activities within the NNI addressing the environmental and health implications of nanotechnology have largely been coordinated through the Nanotechnology Environmental and Health Implications (NEHI) working group. The working group aims to support federal activities to protect public health and the environment through exchanging information across federal agencies, facilitating the identification, prioritisation and implementation of research and promoting the communication of information on the environmental and health impact of nanotechnology to other government and non-government groups. Agencies participating in NEHI include risk-based research and regulatory agencies such as the National Institute for Occupational Safety and Health (NIOSH), the Environmental Protection Agency (EPA), the Food and Drug Administration (FDA) and the Consumer Product Safety Commission (CPSC). A number of other agencies with research capabilities in addressing potential risk, or an interest in risk management, also participate, including the Department of Energy (DOE), the Department of Defense (DOD), the National Institutes of Health (NIH) and the National Science Foundation (NSF).

3   Federal Government Activities in Support of "Safe" Nanotechnology

While the NEHI has provided an effective forum for government agencies to exchange notes and experiences, there has been little direct output visible from the working group since its inception. In the light of the NEHI's coordinating function, much of the group's work is reflected in agency plans and initiatives. Certainly, a number of US federal agencies have developed a position on nanotechnology and have instituted internal nanotechnology programs since joining NEHI. What the working group has contributed to directing strategic risk-related research across the federal government is harder to judge. The first overt step in this direction was anticipated later in 2006, when the NEHI planned to release a report on strategic research needs and directions.[†]

One area that NEHI has contributed to is collating information on nano-technology risk-based research funding levels. As reported in the 2006 NNI supplement to the President's budget,[16] federal agencies were planning to spend $38.5 million on research into the environmental, safety and health impact of nanotechnology between 2005 and 2006 (Table 1). Over 60% of this funding resides in agencies focused on basic research, with approximately 25% being

---

[†] In September 2006, NSET published the report "Environmental, health and safety research needs for engineered nanomaterials"

Table 1 Estimated US annual investment in research and development with relevance to the environmental, safety and health implications of engineered nanomaterials (in millions of dollars). Comparing estimated from the National Nanotechnology Initiative,[a] and the Project on Emerging Nanotechnologies (PEN)[b].

Agency	NNI-estimated investment 2005/2006	PEN-estimated investment, 2005 (any relevance)	PEN-estimated investment, 2005 (highly relevant)
NSF	24.0	19.0	2.5
DOD	1.0	1.1	1.1
DOE	0.5	0.3	0
HHS (NIH)	3.0	3.0[c]	3.0[c]
DOC (NIST)	0.9	1.0	0
USDA	0.5	0.5	0
EPA	4.0	2.6	2.3
HHS (NIOSH)	3.1	3.1[d]	1.9[e]
DOJ	1.5	0	0
Totals	38.5	30.6	10.8

[a]NSET, The National Nanotechnology Initiative, Research and development leading to a revolution in technology and industry. Supplement to the President's FY2006 budget, Nanoscale Science Engineering and Technology subcommittee of the NSTC, 2005.
[b]PEN, Inventory of research on the environmental, health and safety implications of nanotechnology, Project on Emerging Nanotechnologies, Woodrow Wilson International Center for Scholars, 2005.
[c]Estimate, based on research within the National Toxicology Program.
[d]Based on aggregated funding. Reported by NNI.
[e]Estimated from the percentage of projects highly relevant to engineered nanomaterials.

associated with agencies directly addressing risk evaluation and management. These figures are somewhat lower than initial estimates of $100 million per year[19], which included research having incidental relevance to risk.

Three agencies within this list have coordinated research programs addressing the health, safety and environmental implications of nanotechnology: NIH, NIOSH and EPA. Risk-related research within NIH is predominantly administered within the National Institute of Environmental Health Sciences (NIEHS), which oversees the national Toxicology Program (NTP) – a collaboration between NIEHS, NIOSH and FDA.[20] In 2003, a group of nanoscale materials was nominated to the NTP for testing. Research is currently underway within the NTP Nanotechnology Safety Initiative to address the potential human hazards associated with the manufacture and use of nanoscale materials. The intent is to conduct studies that test hypotheses focused on the relationship of key physico-chemical parameters of selected nanomaterials on their toxicity. These currently include the dermal toxicity of materials such as titanium dioxide and zinc oxide, the pulmonary toxicity of single-walled carbon nanotubes and systematic studies of the toxicity of quantum dots, fullerenes and related compounds.

The National Institute for Occupational Safety and Health has been active in disseminating information on nanotechnology in the workplace and it has been actively addressing research and information needs for some years. Current research within the agency that has some relevance to nanotechnology is estimated at approximately \$3 million per year.[16] This research covers the toxicity and health impact of nanomaterials (including carbon nanotubes), exposure evaluation, exposure control and good working practices. In 2005, the agency published a draft strategic research plan for nanotechnology in the workplace that outlined current needs and the agency's plans to address those needs.[21] At the same time, the agency published a draft document entitled "Approaches to safe nanotechnology. An information exchange with NIOSH".[22] This document outlines many of the concerns over engineered nanomaterials in the workplace and the current state of knowledge regarding potential risk and risk assessment/reduction. The National Institute for Occupational Safety and Health has also demonstrated a commitment to studying exposure to engineered nanomaterials in the workplace, and at the end of 2005 the agency announced a program of field studies to be conducted in partnership with industry.[23]

The Environmental Protection Agency started to fund extramural research into the environmental applications and implications of nanotechnology in 2001. Under the Office of Research and Development Science To Achieve Results (STAR) program, the agency has awarded over 32 grants addressing the environmental applications and implications of nanotechnology, worth over \$10 million.[24] In recent years, EPA has been partnering with agencies such as NIOSH, NSF and NIEHS to increase the scope and extent of this research program. As of 2005, estimated funding within EPA into the environmental applications and implications of nanotechnology was \$4 million per year.[16] In 2006, a further \$4 million per year was requested by EPA to support intramural research into nanotechnology and the environment.[25] A draft white paper was published by EPA in December 2005, which provides an idea of the issues the agency considers important and which may receive attention through this new funding.[24]

Gaining further insight into the nature and extent of the research represented in Table 1 is not easy, as the NNI does not release information on specific research projects. One reason cited for not releasing specific information is the difficulty and complexity in identifying research that might have some relevance to risk, and the danger of either over- or under-estimating the extent to which relevant research is being funded. However, information on what is and is not being done is clearly essential to strategic research planning, especially if future risk-related research is to target critical knowledge gaps. In 2005, the Wilson Center Project on Emerging Nanotechnologies (PEN) sought to clarify the research landscape by compiling a publicly accessible on-line inventory of current risk-related research relevant to nanotechnology.[26] Published information on federally funded research was classified by relevance to an understanding of risk, category of nanomaterials (engineered, incidental or natural) and impact sector (human health, environment or safety). This enabled a sophisticated analysis of current research of relevance to the implications of engineered nanomaterials. It also allowed research trends to be explored, and research gaps to be identified.

In Table 1, estimates of annual federal funding for nanotechnology environmental, safety and health research from the PEN analysis are compared with the NNI figures. The comparison is confounded by slightly different reporting periods and a reticence within some agencies to provide detailed information on current research. However, the figures provide a reasonable indication of current activity. Research with some relevance to risk includes research into nanotechnology applications which might also be relevant to understanding health, safety and environmental impacts, while highly relevant research only includes projects specifically focused on understanding risk. Reported funding of research with some relevance to risk matches NNI figures reasonably well. However, identified funding for research that is highly relevant to risk is only $11 million per year – less than 1% of the annual US nanotechnology R&D budget. Although not conclusive, comparison of the NNI and PEN figures suggests only one third to one quarter of reported NNI funding into the environmental safety and health implications of nanotechnology has a high degree of relevance to these specific risk issues.

Although funding levels are a useful indicator of activity, an analysis of how these funds are being used provides greater insight into the relevance of current research. Research listed in the PEN inventory indicates that there is little to no strategic direction to risk-based research in the USA. For example, an analysis of research into impact according to different exposure routes indicated a heavy emphasis on inhalation and a negligible emphasis on ingestion – even though inhaled and intentionally eaten nanomaterials will enter the gastrointestinal tract.[27]

The concern that engineered nanomaterials might behave differently from conventional materials has sparked debate over the applicability of oversight and regulatory mechanisms in the USA. The established position of the US government is that the current regulatory framework is sufficiently robust to accommodate emerging nanotechnologies and engineered nanomaterials, although some changes at an operational level may be required.[28] However, individual regulatory agencies are beginning to consider their response to emerging nanotechnologies, engineered nanomaterials and nano-enabled products. The Environmental Protection Agency began development of a voluntary program for industry in 2005,[24] which would require participants to provide existing information and generate new information on the potential health and environmental impact of engineered nanomaterials. The Food and Drug Administration held a public meeting in October 2006 to gather information about current developments in uses of nanotechnology materials regulated products. The Consumer Product Safety Commission released a statement in 2005 outlining the agency's mission and authority and highlighted some of the challenges facing the regulation of nanotechnology-based products.[29]

An independent report published in 2006 by J. Clarence Davies calls into question the robustness of current regulatory frameworks in the USA when applied to nanotechnology.[4] Written with the aim of stimulating dialogue at a critical time for nanotechnology, Davies cites the inability of chemical or physical properties alone to predict the behavior of nanomaterials as a rationale for considering new nano-specific regulation. He concludes that nanotechnology is difficult to address using existing regulations, that a new law may be required to manage potential risk and that new mechanisms and institutional capabilities are required.

While it is still uncertain how the oversight of nanotechnology will develop within the USA, it remains a focus of serious consideration within the federal government. Hearings of the House Science Committee in November 2005 and the Senate Committee on Commerce, Science and Transportation in February and May 2006 have all addressed nanotechnology oversight, and further hearings addressing the implications of nanotechnology are planned.

4   Industry and Other Non-government Activities in Support of "Safe" Nanotechnology

The non-government community in the USA continues to be active in addressing the safety of nanotechnology. Some activities – such as the development of standards, information dissemination and research coordination – are proceeding within an international context. Others are more focused on what is happening within the USA, but which will also have an international impact. In general it is clear that different stakeholders recognise a need to ensure the risks of new nanotechnologies are minimised.

A 2006 report from the RAND Corporation on nanomaterials in the workplace synthesised the perspectives of many nanotechnology stakeholders, concluding that "public and private resources and funds being allocated to understanding the occupational, health, and environmental risks of emerging nanomaterials are not commensurate with the development of new nanomaterials".[30] The same report underlines the need to address risk in a coordinated and strategic manner if nanotechnology-based enterprises are to succeed. Industry-led groups, such as the Chemical Industry/Semiconductor Industry Consultative Board on Advancing Nanotechnology, have outlined research gaps that need to be filled in support of "safe" nanotechnology.[31] Environmental Defense – a non-profit organisation seeking scientifically sound and sustainable solutions to environmental issues – has called for substantially increased and sustained government support of environmental, safety and health research and development in the field of nanotechnology.[32]

The group has also been working with industry to develop sustainable solutions to "safe" nanotechnology. For instance, in October 2005 they announced a collaboration with DuPont to develop a framework for the "responsible development, production, use and disposal of nano-scale materials". DuPont is also actively supporting research into assessing and managing potential risks associated with engineered nanomaterials. As well as a productive in-house research program, they are leading an industry-based research collaboration to develop a better understanding of the behaviour of airborne nanostructured particles, and how to measure and prevent exposure effectively.[33] Further non-government-led initiatives are emerging within multi-stakeholder groups such as the International Council On Nanotechnology (ICON). Bringing government and non-government stakeholders together, ICON is active in addressing relevant risk-related issues within the context of a broad community. In 2005 the Council published a web-based database of peer review publications relevant to nanotechnology and risk, with the intention of summarising and synthesising information into an easily accessible knowledge base.[‡] Research supported by ICON is currently examining good practices for working with engineered nanomaterials, including what is currently being done and what needs to be done.

ICON members and others are also closely involved in nanotechnology standards development. While standards development is occurring at an international level, it is currently playing an important role in addressing "safe" nanotechnology within the USA. In 2005 the International Standards Organization (ISO) formally established Technical Committee 229 to address nanotechnology standards. This Technical Committee currently has three working groups addressing terminology and nomenclature, measurement and characterisation and health, safety and environmental aspects (the latter is being coordinated from the USA through the American National Standards Institute). In parallel with this effort, ASTM International Technical Committee E56 is developing international nanotechnology standards that include management of environmental occupational health and safety risk, and product stewardship. Prior to either technical committee being established, ISO TC146 began work on a Technical Report addressing the measurement of airborne exposure to nanostructured particles in the workplace. This report was approved for publication early in 2006.[34]

Taking a broader perspective on nanotechnology and science policy within the USA, the Project on Emerging Nanotechnologies was established in 2005 with the aim of bringing stakeholders together in a dialogue to develop sound policy, relevant research and safe practices. A partnership between the Woodrow Wilson International Center for Scholars (established in 1968 by Congress as a non-partisan living memorial to President Woodrow Wilson) and the Pew Charitable Trusts, the Project has been influential in raising awareness of the potential benefits and risks of emerging nanotechnologies and in enabling a broad dialogue on research and policies to underpin sustainable nanotechnologies. In 2006, the Project released the first publicly available on-line inventory of nanotechnology-based consumer products, to inform people about how this technology is entering their lives, and to support informed nanotechnology risk research and policy decisions (Figure 1). The Project on Emerging Nanotechnologies is currently one of the most widely cited sources of information in the USA on the responsible development and implementation of nanotechnology.

5 Looking to the Future – Ensuring the Development of "Safe" Nanotechnology

Reviewing current US activities in support of "safe" nanotechnology, it is clear that there is recognition of the need to address risk and a willingness to act to a

---

[‡] http://icon.rice.edu/research.cfm.

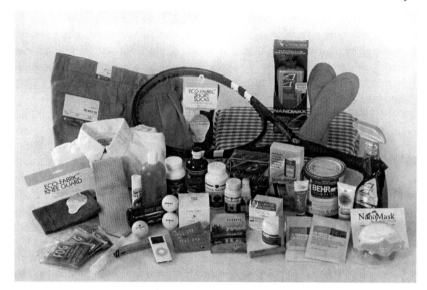

Figure 1   A cross-section of manufacturer-identified nanotechnology consumer products available now. An on-line inventory hosted by the Wilson Center Project on Emerging Nanotechnologies lists over 200 products globally[§].[40] © 2006 David Hawkshurst/Woodrow Wilson International Center for Scholars.

certain degree on this recognition. However, there remains a dearth of information on risk and risk management. An analysis of nanotechnology risk research gaps from diverse organisations and sectors highlights how little is still known about the impact of the technology on human health and the environment and how much more research is needed.[3,21,24,31,32,35–38] The cross-disciplinary nature of nanotechnology requires risk-based research that transcends traditional boundaries. This in itself challenges conventional ways of doing science and suggests the need for an overarching strategic research framework. In addition, risk-based research must ultimately be applied to ensuring that people and the environment are not harmed, and this requires a close association between research and oversight. Finally, sufficient resources are required to carry out effective research, including facilities, personnel and research funds. An examination of current US activities suggests little or no strategic coordination, no clear link between research and oversight and insufficient resources to make a significant difference. While it is perhaps disingenuous to surmise that the current emphasis on getting the environmental safety and health aspects of nanotechnology right is little more than window dressing, it is clear that more needs to be

---

[§] www.nanotechproject.org/consumerproducts.

done if a serious attempt is to be made to understand and minimise risk early on in the technology's development and implementation.

Looking to the future of "safe" nanotechnology within the USA and globally, it is useful to consider what a viable strategic research framework might look like. An effective framework for strategic nanotechnology risk-based research is likely to have a number of attributes. It will provide a link between the implementation of nanotechnologies and the research necessary to ensure appropriate oversight of risk; it will ensure coordinated direction of research within different agencies and organisations at a national level; it will enable coordination and partnerships between international initiatives; it will allow resources to be allocated appropriately to address critical issues; and it will provide broad strategic research priorities for assessing and managing potential risk. A successful research framework that underpins sustainable nanotechnologies will also be responsive to the increasing sophistication of these technologies, and will evaluate progress against needs through review and revision.

Who should be responsible for such an overarching framework? Industry stands to gain a lot from nanotechnology according to some sources.[39] It is certainly in industry's best interest to ensure that appropriate strategic research frameworks are put in place in order to maintain public and commercial confidence in their products, as well as to minimise the chances of adverse impacts. However the question that must be addressed first and foremost is: What is in the best interest of the society and environment in which we live? Conceptually, this does not seem an appropriate question for industry to take the lead on. Perhaps more pragmatically, while industry has been shown to be more than capable of directing and funding research that addresses product-specific risk, it is difficult to find an economic justification for having industry lead in developing a basic understanding of risk.

The most viable alternative to an industry-led strategic research framework is a government-led framework. A strategic research framework developed and administered by the government would combine societal accountability with a high-level overview of research needs, a capacity for addressing generic and applied issues and a facility for partnerships and coordination. It can also be argued that the federal government has a social responsibility for developing and implementing an effective strategic research framework. The US federal government is investing a lot of money into nanotechnology research and development – over $1 billion dollars a year.[16] With this investment comes a certain degree of social responsibility – to ensure new risks associated with resulting technologies are assessed and managed appropriately. This is a responsibility to people who may be directly or indirectly affected by new risks. It is also a responsibility to the business community, who need to know the social and technical risks associated with the technologies they are being encouraged to develop.

Assuming that the US government were to develop and implement a strategic research framework addressing the environmental, safety and health implications of nanotechnology, the central pillars of a workable strategy would most likely include:

Linking research to oversight. Ultimately, the aim of a strategic risk-related research framework will be to minimise and manage risk through applying existing knowledge and developing new knowledge. However, this work will be ineffective in the long term if research is not linked to oversight, whether this takes the form of regulation, voluntary programs, best practices or other risk management tools and approaches.

Balancing basic and applied research. Answers to short-term critical research questions require applied research, while understanding mechanisms of risk and risk management must be underpinned by basic (or pure) research. Both modes of research have their place. However, an effective strategic research framework will ensure that the types and models of research are employed to match real-world research needs.

Authority to direct and support research. An effective strategic research framework must have teeth. It will not be sufficient merely to suggest areas of research to respective agencies or to rely on agency resources to support the necessary research. While a certain level of autonomy must be directed to research organisations, an effective strategic research framework will include mechanisms that ensure work is done by the appropriate organisations, and that resource levels are adequate to the task.

Coordination and partnership. As well as directing and coordinating research within the federal government, a strategic research framework will only be successful if it includes provisions to coordinate and partner with industry, international governments and non-government organisations. With such provisions, international resources – both private and public – may be brought to bear with maximum effect and minimum redundancy in managing new risks associated with emerging nanotechnologies.

Whether such a strategic framework will emerge within the USA, or indeed globally, is not yet clear. However, it probably is not too great an exaggeration to say that the long-term health of nano-businesses, as well as the people they employ and who use their products, will depend on well-funded and appropriately directed research into understanding and minimising the risks of emerging nanotechnologies.

Update

Nanotechnology is a rapidly developing field and inevitably, any overview of activities is out of date almost before it is completed. Since this chapter was written, industry, government and other organizations based in the US have made significant strides to frame the challenges and opportunities being faced. Notable highlights include a list of environmental, safety and health research needs published by the NSET subcommittee in September 2006,[41] publication of the EPA white paper on nanotechnology in early 2007,[42] and publication of the DuPont/Environmental Defense Nano Risk Framework in 2007.[43] These and other activities are helping to clarify what needs to be done to ensure the success of nanotechnologies through understanding and avoiding potential risks. However, strategic action to ensure the safe use of engineered nanoparticles remains slow in coming. A report from the Project on Emerging Nanotechnologies published in July 2006 outlined a potential research strategy for the US to address short, medium and long-term challenges.[44] This was complemented by a paper published in Nature later in 2006, co-authored by 14 internationally recognized scientists, outlining five "grand challenges" to addressing potential nanotechnology risks.[45] In March 2007, the UK Center for Science and Technology urged the UK government to take "swift and determined action necessary to regain its leading position in nanotechnologies".[46] This is a message that seems as relevant to the United States as it is to any country hoping to reap the potential benefits of engineering particles at the nanoscale.

References

1. ETC Group, No Small Matter II: The Case for a Global Moratorium. Size Matters! ETC Group, 2003.
2. E. Hood, *Environ. Health Perspec.*, 2004, **112**, A741–A749.
3. The Royal Society and The Royal Academy of Engineering, Nanoscience and nanotechnologies, The Royal Society and The Royal Academy of Engineering, 2004.
4. J.C. Davies, Managing the effects of nanotechnology, Woodrow Wilson International Center for Scholars, Project on Emerging Nanotechnologies, 2006.
5. A. Hett, Nanotechnology. Small matter, many unknowns, Swiss Re, 2004.
6. O. Renn, Risk Governance. Towards an integrative approach, International Risk Governance Council, 2005.
7. J. McCoubrie, Informed public perceptions of nanotechnology and trust in government, Woodrow Wilson International Center for Scholars, Project on Emerging Nanotechnologies, 2005.
8. M.C. Roco, *J. Nanopart. Res.*, 2003, **5**, 181–189.
9. J. Van, Nanotechnology industry puts focus on safety issues, Chicago Tribune, p. 3, Chicago, January 21 2006.
10. C. Arthur, Does Scarlett need regulatory oversight?, The Guardian, London, January 19 2006.
11. R. Weiss, Nanotechnology Precaution Is Urged. Minuscule Particles in Cosmetics May Pose Health Risk, British Scientists Say, Washington Post, p. A02, Washington DC, July 30 2004.
12. T. Hampton, *J. Am. Med. Assoc.*, 2005, p. 2564.
13. G. Oberdörster, E. Oberdörster and J. Oberdörster, *Environ. Health Perspect.*, 2005, **13**, 823–840.
14. National Academies, Small wonders, endless frontiers. A review of the National Nanotechnology Initiative, National Academy Press, 2002.
15. IWGN, Nanotechnology Research Directions: IWGN Workshop Report. Vision for Nanotechnology R&D in the Next Decade, National Science

and Technology Council Committee on Technology Interagency Working Group on Nanoscience, Engineering and Technology (IWGN), 1999.

16. NSET, The National Nanotechnology Initiative. Research and development leading to a revolution in technology and industry. Supplement to the President's FY2006 budget, Nanoscale Science Engineering and Technology subcommittee of the NSTC, 2005.

17. US Congress, 21st Century Nanotechnology Research and Development Act (Public Law 108–153), 108th Congress, 1st session, 2003.

18. NSET, The National Nanotechnology Initiative Strategic Plan, National Science and Technology Council, 2004.

19. C. Stuart, *Small Times*, 2006, **6**, 22–23.

20. NTP, National Toxicology Program. Current directions and evolving strategies, National Institute of Environmental Health Sciences, NIH, 2006.

21. NIOSH, Strategic plan for NIOSH nanotechnology research. Draft, September 28 2005, NIOSH, 2005.

22. NIOSH, Approaches to safe nanotechnology. An information exchange with NIOSH, National Institute for Occupational Safety and Health, 2005.

23. NIOSH, NIOSH to Form Field Research Team for Partnerships in Studying, Assessing Nanotechnology Processes, Web Address: www.cdc.gov/niosh/topics/nanotech/newsarchive.html#fieldteam, accessed May 24 2006.

24. EPA, U.S. Environmental Protection Agency Nanotechnology White Paper: External Review Draft, EPA, 2005.

25. OSTP, National Nanotechnology Initiative. Research and development funding in the President's 2007 budget, Office of Science and Technology Policy, 2006.

26. PEN, Inventory of research on the environmental, health and safety implications of nanotechnology, Project on Emerging Nanotechnologies, Woodrow Wilson International Center for Scholars, 2005.

27. A.D. Maynard, *Nano Today*, 2006, **1**, 22–33.

28. S.R. Morrissey, *Chem. Eng. News*, 2006, **84**, 34–35.

29. CPSC, CPSC Nanomaterial Statement, Consumer Products Safety Commission, 2005.

30. J.T. Bartis and E. Landree, Nanomaterials in the workplace. Policy and planning workshop on occupational safety and health, The RAND Corporation, 2006.

31. Chemical Industry Vision 2020 Technology Partnership; SRC Joint NNI-ChI CBAN and SRC CWG5 Nanotechnology research needs recommendations, 2005.

32. R.A. Dennison, A proposal to increase federal funding of nanotechnology risk research to at least $100 million annually, Environmental Defense, 2005.

33. K. Doraiswamy, Statement of Krishna Doriaswamy, Ph.D. Research Planning Manager, DuPont Central Research & Development, before the Committee on Science, U.S. House of Representatives, November 17 2005, DuPont, 2005.

34. ISO, Workplace atmospheres - Ultrafine, nanoparticle and nano-structured aerosols – Exposure characterization and assessment, International Standards Organization ISO/TR 27628, 2006.
35. G. Oberdörster, A. Maynard, K. Donaldson, V. Castranova, J. Fitzpatrick, K. Ausman, J. Carter, B. Karn, W. Kreyling, D. Lai, S. Olin, N. Monteiro-Riviere, D. Warheit, and H. Yang, *Part. Fiber Toxicol.*, 2005, **2**, doi:10.1186/1743–8977–1182–1188.
36. A.D. Maynard, and E. D. Kuempel, *J. Nanopart. Res.*, 2005, **7**, 587–614.
37. HM Government, Characterizing the potential risks posed by engineered nanoparticles. A first UK government research report, Department for Environment, Food and Rural Affairs, 2005.
38. EC, Communication from the commission to the council, the European parliament and the economic and social committee. Nanoscience and nanotechnologies: An action plan for Europe 2005–2009, Commission of the European Communities, 2005.
39. Lux Research, Sizing nanotechnology's value chain, Lux Research Inc., 2004.
40. PEN, The nanotechnology consumer products inventory, Project on Emerging Nanotechnologies, Woodrow Wilson International Center for Scholars, 2006.
41. NSET, *Environmental, health and safety research needs for engineered nanoscale materials*; Subcommittee on Nanoscale Science, Engineering and Technology, Committee on Technology, National Science and Technology Council: Washington DC, 2006.
42. EPA, US Environmental Protection Agency Nanotechnology White Paper. In EPA, Ed. 2007.
43. DuPont; Environmental Defense, *Nano Risk Framework*; DuPont and Environmental Defense: 2007.
44. A. D. Maynard, *Nanotechnology: A research strategy for addressing risk*; PEN 03; Woodrow Wilson International Center for Scholars, Project on Emerging Nanotechnologies: Washington DC, 2006.
45. A. D. Maynard, R. J. Aitken, T. Butz, V. Colvin, K. Donaldson, G. Oberdörster, M. A. Philbert, J. Ryan, A. Seaton, V. Stone, S. S. Tinkle, L. Tran, N. J. Walker and D. B. Warheit, Safe handling of nanotechnology, *Nature*, 2006, **444**(16), 267–269.
46. CST, *Nanosciences and nanotechnologies: A review of government's progress on its policy commitments*; Council for Science and Technology: London, UK, 2007.

Subject Index